TRANSDUCER
INTERFACING
HANDBOOK

Analog Devices Technical Handbooks

Sheingold, ed.: Analog-Digital Conversion Handbook, 1972
    (Out of Print)
Sheingold, ed.: Nonlinear Circuits Handbook, 1974
Burton and Dexter: Microprocessor Systems Handbook, 1977
Sheingold, ed.: Analog-Digital Conversion Notes, 1977
Sheingold, ed.: Transducer Interfacing Handbook, 1980
Boyes, ed.: Synchro and Resolver Conversion, 1980

TRANSDUCER
INTERFACING
HANDBOOK

A GUIDE TO
ANALOG SIGNAL CONDITIONING

Edited By
Daniel H. Sheingold

Published by
Analog Devices, Inc.
Norwood, Massachusetts 02062 U.S.A.

ISBN 0-916550-05-2
Library of Congress Catalog Card No. — 80-65520

Copyright© 1980, 1981 by Analog Devices, Inc.
Printed in the United States of America
First edition, March, 1980
3 4 5 6 7 8

Additional copies may be ordered from Analog Devices, Inc., P.O. Box 796, Norwood, Mass. 02062

PREFACE

In a continuing trend, there is an increasing use of measurement and control in Industry to improve efficiencies and reduce costs: environmental, economic, and energy-related. The *transducer* is an essential link in such a process. This book is about the electrical aspects of commonly used transducers that sense temperature, pressure, force, level, and flow.

Input transducers interface with electrical systems to provide electrical outputs that are indicative of the state of the phenomena being sensed. Usually, certain kinds of electrical interfacing processes are required, in order to provide analog information that is conveniently usable by the system. Such *signal conditioning* includes providing excitation, amplifying, filtering, linearizing, isolating, and offsetting.

This book is about the interfacing of transducers to electrical analog circuitry in preparation for readout, further analog transmission or processing, or conversion to digital form. It deals with the signal conditioning required to excite some of the most commonly used physical-to-electrical (*input*) transducers, e.g., thermistors, RTD's, strain gages, and to amplify and normalize their outputs. The book's objective is to fill the information gap that may exist between competent specialists on both sides of the interface.

Who was the book written for? One reader who might benefit is the electrical systems engineer, with a solid background in digital hardware and software, who must apply this knowledge to the measurement and control of physical phenomena via the inevitable analog signals that they spawn. Another is the mechanical engineer or physicist, who has a deep knowledge of the systems (s)he is measuring and controlling, enough electronics and good sense *not* to want to become an expert electronic designer, and a desire to know some practical options for a solution to the instrumentation problem. Then there are: those who are somewhere between the two poles; students with expectations of some day having similar needs (in thesis projects or in their working careers); the instructors who teach these students; and—finally—anyone interested in learning about practical approaches to the electrical aspects of solving measurement problems.

The text provides technical information essential for an informed "make-or-buy" electrical-interface-circuit decision. In the process, it describes nearly one hundred applications including some specialized complete measurement-and-control systems that have been designed and built using basic interface circuitry and standard transducers. The viewpoint throughout is that of the practical scientist, engineer, or technician; profligate use of mathematics is avoided.

We have intentionally limited the scope of the book. In order for a book on transducers to do full justice to the subject, it would have to be at least an order of magnitude larger than the present volume; yet ninety percent of the information it contained would be of use to only ten percent of its readers. When planning this book, we decided to substitute for a doctrinnaire thoroughness a useful *concentration* on those topics that have the strongest economic impact on the largest body of present and potential transducer users. The term "economic" refers to costs and benefits, measured not only in money, but also in the *time* of the busy and hard-pressed systems engineer.

Consequently, this book is principally about commonly used transducers that sense *temperature*, *pressure*, *force*, *level*, and *flow*. A large number of eminently worthy, useful, and occasionally fascinating phenomena, and their related sensors, which do not have as widespread application, have (sometimes regretfully) been omitted; examples of these include photometry and acoustic and infra-red thermometry. We have also chosen not to include position and velocity, even though our line of resolver- and synchro-digital conversion interface products is uppermost in our thoughts for many such applications.*

In addition, we have chosen to avoid discussing certain specialized transducers that are almost always purchased with matching interface electronics, usually furnished by a transducer's manufacturer. Examples include linear variable differential transformers (LVDT's), pH probes, and some capacitive transducers. In such cases, information of the kind provided here would be at once insufficient and (paradoxically) superfluous.

Finally, this book avoids the detailed consideration of interfacing in the *physical* realm. Such important particulars are in many cases an integral part of the user's expertise, and they are often

---

*For in-depth background on synchros and resolvers in positional measurements, the 208-page book, *Synchro and Resolver Conversion*, edited by Geoff Boyes, has just been published (1980) and is available from Analog Devices at reasonable cost.

addressed in substantial detail by sensor manufacturers. Our concentration on the *electrical* interface is in harmony with our specialization in the manufacture of electronic products for precision measurement and control. It is in this area that our readers receive the most benefit—benefits that we share, and the best of reasons for producing such a book as this

ACKNOWLEDGEMENTS

The idea of publishing a handbook on signal conditioning for transducers was first suggested by Jim Williams. This book draws heavily from a manuscript that he wrote for us, based on his extensive experience in marrying electronic circuits with physical phenomena and his copious files of working measurement-and-control circuits that use transducers.

A great many technical and strategic ideas, an endless flow of transducer product information, and helpful chapter-by-chapter review and critique were provided by Frank Goodenough. Sanction for the book and a number of cogent suggestions that provided a basis for making the difficult decisions as to what to include and what to leave out, as well as encouragement when the going got sticky, were furnished by Bob Boole.

Contributions on technical matters came from every quarter of Analog Devices, both directly and via the already documented contributions in our literature. Though it is difficult to single out specific contributors, valuable comments directly affecting the content and organization of the book were received from George Reichenbacher and Janusz Kobel. Lew Counts and Jeff Riskin have reviewed the entire manuscript and provided valuable (sometimes pungent) comments. Portions of the book have been reviewed by Mike Timko, Jim Maxwell, and others.

By the time this book has been published, some of the chapters may have appeared in the trade press. Eric Janson will have been the interface, performing any needed excitation and signal conditioning.

The book was produced in our Publications Department, under the direction of Marie Etchells. Kathy Hurd and Joan Costa composed the type, Ernie Lehtonen, Dianne Nemiccolo, and Jean

Ellard did the drawings and mechanical layouts, and it was printed at Banta Press. The book is made available through our Literature Fulfillment and Distribution group, under Cammy O'Brien.

Much patience was asked of and unstintingly furnished by my wife, Ann, and my children.

If the book (the first of its kind that we are aware of) turns out to be as valuable a contribution as we had hoped, all of the above deserve the major share of the recognition.

Daniel H. Sheingold

Norwood, Mass.
January 2, 1980

TABLE OF CONTENTS

The Transducer as a Circuit Element

Chapter 1

TRANSDUCERS – ACTIVE AND PASSIVE

Webster's New Collegiate Dictionary defines *transducer* as "a device that is actuated by power from one system and supplies power, usually in another form, to a second system." In this book we are concerned with *input transducers*–actuated by physical variables representing force, pressure, temperature, flow, and level–that supply electrical signals to the front end of a measurement and control system (which may range in complexity from a simple analog meter to a multi-input, multi-output, multiprocessor, multi-loop digitally controlled system).*

The input transducers with which we are concerned may be categorized in a number of ways. Number of energy ports, input variable, sensing element, and electrical circuit configuration are some specific forms of differentiation that will be treated here.

From the standpoint of energy, there are two broad classes, *active* and *passive* transducers.† A *passive*, or self-generating, transducer is one which has an input and an output (i.e., two *energy ports*). All of the electrical energy at the output is derived from the physical input; examples include thermocouples, crystal microphones, and photodiodes in the photovoltaic mode. Since the electrical output is limited by the physical input, such transducers tend to have low-energy outputs requiring amplification, but this is not always true: a piezoelectric accelerometer may furnish *watts* of

*Input transducers may be contrasted with *output transducers*, which convert electrical to other forms of energy, e.g., loudspeakers, solenoids, etc.

†These terms are restricted cases of the more-general *active* and *passive* circuits. (*IEEE Standard Dictionary of Electrical and Electronics Terms*, 2nd Edition, Institute of Electrical and Electronics Engineers, Inc., New York, 1977)

power for short periods during explosions or hammer blows.

An *active* transducer has a physical input, an electrical output and an electrical *excitation* input (i.e., three energy ports). The physical input, in effect, modulates the excitation. Examples of this type of transducer include resistance strain gages* and bridges, platinum resistance thermometers, and semiconductor temperature sensors. Active transducers are usually of the *open-loop* type with fixed excitation. However, there is a subclass of active transducers in which the electrical output serves as an indication of unbalance (or *error*), and the excitation or the physical parameter is manipulated in a feedback configuration to maintain balance; the feedback signal is then a measure of the physical variable. Some members of this class are known (appropriately) as *force balance* transducers.

To the interface designer, one significance of these two categories is that passive transducers must be dealt with strictly on their own terms. The form of response—and its sensitivity—are limited by the energy available in the phenomenon and the device's conversion efficiency; very little negotiating, cajoling, or finagling is possible. For example, with thermocouples, we are stuck with their low sensitivity, low output level, and nonlinear response. While we can choose different combinations of metals to produce different outputs over varying temperature ranges, we cannot budge the characteristics of a given thermocouple. On the other hand, their structure is simple, their characteristics are consistent, interchangeable, and reliable (when properly implemented), and standard output tables are published for the various couplings.†

With active transducers, there is an additional degree of freedom. Although they rely on fundamental properties of materials in the same way that passive devices do, the excitation can be used (in many instances, but by no means always) to provide an increased output level, but there are tradeoffs.

Consider a simple resistance temperature-detector circuit (Figure 1-1). The current source, I, develops across resistance R—which

---

*It can be argued (correctly) that the proper spelling of this word is "gauges". However, through consistent usage in the industry, the simplified spelling—when used in the term "strain gage"—has become a *de facto* standard and will be so employed throughout this book when used in that context.

†Increased sensitivity can be gained by connecting sets of junctions in series—physically associating alternate junctions—to form *thermopiles*, but their cost and complexity restrict their applicability.

varies predictably with temperature—a small voltage. For accurate resolution of small temperature changes, with correspondingly small resistance changes, it would be useful to increase I. For example, if I were increased tenfold, the resolution would increase in the same degree. However, a tenfold increase in I produces a hundredfold increase in dissipation. When I is sufficiently great, the power dissipated in R will itself cause the temperature to rise perceptibly, introducing a measurement error.

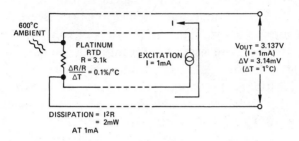

a. Constant-current excitation, I = 1mA

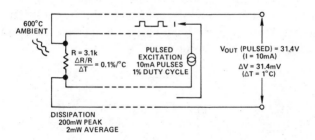

b. Pulsed excitation, 10mA rectangular pulses

Figure 1-1. Active transducers may provide increased output if the excitation is increased in an appropriate way

A possible solution to this problem would be to *pulse* the current at 10I, using accurate rectangular pulses with a duty cycle of 1% at a frequency that is high compared to the thermal response of the device. The average dissipation would be the same as for continuous I, but the sampled pulses would provide a tenfold improvement in sensitivity. This example is mentioned here, not as a specific technique, but rather as a demonstration of the additional degree of freedom available to the designer for obtaining improved performance via modifications to the circuit parameters of the active transducer's circuitry.*

INTERFACING AND SIGNAL CONDITIONING

Once a particular form of transducer has been chosen—or mandated—for a given job, provisions must be made for appropriate excitation and for conditioning the output signal. The nature of the conditioning depends on the electrical characteristics of the transducer and the destination of the signal. Typical processes include galvanic isolation, impedance transformation, level translation, amplification, linearization, and a variety of computations (analog or digital). These processes may occur in the vicinity of the transducer, at a remote data-acquisition subsystem, or piecemeal at several locations. A common initial form of conditioning is to amplify the signal to a standard voltage range for data acquisition (e.g., 0 to +10V); another is to translate it to a standard process-control current range (e.g., 4 to 20mA) for transmission over a twisted pair to a remote destination. The equipment used for signal conditioning might range from user-designed equipment using electronic components, such as op amps and instrumentation amplifiers, to packaged signal-conditioning modules, to intelligent data-acquisition subsystems.†

Whatever form the conditioning takes, however, the circuitry and performance will be governed by the electrical character of the transducer and its output. Accurate characterization of the transducer in terms of parameters appropriate to the application, e.g., sensitivity, voltage and current levels, linearity, impedances, gain,

---

*Hardened system designers and veteran micro—and milli—volt chasers among our readers will recognize what a prize has been won here; a tenfold increase in output of a transducers is really substantial. On the other hand, although the strategy outlined for this particular example will (and has) worked well, it may well be a last resort: in the case of a strain-gage bridge with 100V pulses in lieu of a continuous 10V, one should bear in mind the possible mechanical effects and contribution to "creep" (see page 20) of a fast-rising 100V pulse applied to the low-impedance bridge. Also, the requirements for circuitry to implement an accurate measurement while minimizing electromagnetic interference are not trivial.

†Analog Devices manufactures a wide variety of useful products for these applications. Examples of products in these categories that the reader may meet later in this volume include the AD521 and AD522 instrumentation amplifiers, Models 261 and AD517 op amps, 2B31 signal conditioners, 2B20 and 2B22 voltage to current-loop transducers, Model 289 isolators, and the unique MACSYM intelligent Measurement And Control sub-SYsteM. Technical data on these and many other products are available from Analog Devices. A variety of other useful publications from Analog Devices are described in the first section of the Bibliography.

offset, drift, time constants, maximum electrical ratings, and stray impedances, as well as any other germane considerations, can spell the difference between substandard and successful application of the device, especially in cases where high resolution and precision, or low-level measurements are involved.

TRANSDUCER CHARACTERISTICS

In this chapter, we will characterize a number of the more-common types of transducer. In the discussions to follow, some devices will be given cursory treatment, while others will be treated in some detail, in the interest of keeping this text to manageable size and serving the major part of the needs of the majority of our readers, an objective expressed in the Preface. For those who desire greater depth, some useful references with fanout are provided in the Bibliography. As noted in the Preface, the quantities to be measured (measurands) that will be discussed in this book are:

Temperature
Force
Pressure
Flow
Level

Temperature Transducers

Temperature is perhaps the most common and fundamental physical parameter an engineer is likely to be called upon to measure. The intimate relationship between processes—physical, chemical, and biological—and temperature, as an index of state, is a primary consideration, from the molecular level to the completed system. In electronics, no other physical phenomenon is as pervasive in its influence on circuitry and systems as is temperature. Consequently, there are a number of phenomena that can be called upon to perform electrical operations as a function of temperature. Those to be discussed here include thermal expansion (bimetallic elements and mercury-column switches), Seebeck voltage (thermocouples), resistance effects (RTD's and thermistors), and semiconductor junction effects (diodes and PTAT* current-output devices, such as the AD590).

The *bimetallic thermal switch* is perhaps the most elementary of

*Proportional To Absolute Temperature

electrical sensors. Devices of this type utilize metals with differing thermal coefficients of expansion to physically make or break an electrical contact at a preset temperature. Familiar examples are the sensors in thermostats used in heating, ventilating, and air-conditioning systems. Disc-shaped bimetallic elements are used to provide snap action (Figure 1-2). They will function from sub-zero temperatures to almost 300°C. Contacts are available in a variety of forms and will handle low-current *dry-circuit* switching as well as currents to beyond 15 amperes. In addition, the hysteresis of the switching point can be specified (for insensitivity to small temperature fluctuations), and the devices are inexpensive.

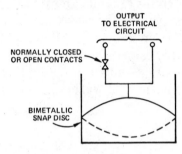

Figure 1-2. Bimetallic disc will snap into position indicated by the dotted lines at the transition temperature

Although the high mass of these units results in a slow thermal response, they are suitable for the many control applications in which temperature tends to change slowly, such as crystal ovens and gyroscopes. They are popular as temperature-limit and override sensors; and they are reliable. Manufacturers* include Elmwood Sensors, Fenwal, Texas Instruments' Klixon Division.

The *electrical-output mercury thermometer* is generically related to the bimetallic thermoswitch, because it relies on differential thermal expansion—in this case, a column of mercury in a glass stem. Fine wires extend into the path of the mercury column, which closes the circuit when the trip point has been reached (Figure 1-3). The trip point can be sharply defined, hysteresis is almost negligible, and life can be quite long (in a physically protected environment). Response time is 1 to 5 seconds (at the trip

*Companies named in these pages are representative manufacturers of the devices with which our transducer people have had the most experience. Mention here should not necessarily imply a recommendation. Buyers' Guides are available with more-complete listings and should be consulted. Some sources of vendor information are to be found in the Bibliography.

point), absolute accuracy to within 0.05°C is available, and multiple—or adjustable—contacts are standard options. These thermometers, with appropriate buffering are excellent choices for control applications and can control a load (that is properly designed) with stability to within 0.01°C. Units of this type can be obtained from P.S.G. Industries.

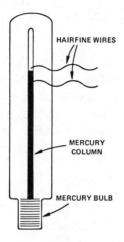

Figure 1-3. The electrical-output mercury thermometer. An electrical contact is established when the mercury column bridges the hairfine wires.

Since the contacts of these thermometers must pass low switching current (typically less than a few milliamperes), transistors, comparators, or sensitive relays can be used to sense the state of the contact closure and provide power for controlling an external device.

Thermocouples are economical and rugged; they have reasonably good long-term stability. Because of their small size, they respond quickly and are good choices where fast response is important. They function over temperature ranges from cryogenics to jet-engine exhaust and have reasonable linearity and accuracy.

Because the number of free electrons in a piece of metal depends on both temperature and composition of the metal, two pieces of dissimilar metal in isothermal contact will exhibit a potential difference that is a repeatable function of temperature. In general, these voltages are small. Table 1-1 lists a number of standard thermocouples, their useful temperature range, and the voltage swing over that range; it can be seen that the average change of voltage with temperature ranges from 7 to about $75\mu V/°C$.

TABLE 1-1. SOME COMMON THERMOCOUPLES

Junction Materials	Typical Useful Temp Range (°C)	Voltage Swing Over Range (mV)	ANSI Designation
Platinum-6% Rhodium — Platinum-30% Rhodium	38 to 1800	13.6	B
Tungsten-5% Rhenium — Tungsten-26% Rhenium	0 to 2300	37.0	(C)
Chromel — Constantan	0 to 982	75.0	E
Iron — Constantan	-184 to 760	50.0	J
Chromel — Alumel	-184 to 1260	56.0	K
Platinum — Platinum-13% Rhodium	0 to 1593	18.7	R
Platinum — Platinum-10% Rhodium	0 to 1538	16.0	S
Copper — Constantan	-184 to 400	26.0	T

Since *every pair* of dissimilar metals in contact constitutes a thermocouple (including copper/solder, about $3\mu V/°C$ and Kovar/rhodium), and since a useful electrical circuit requires at least two contacts in series, measurements with thermocouples must be implemented in a manner which minimizes undesired contributions of incidental thermocouples and provides a suitable reference.

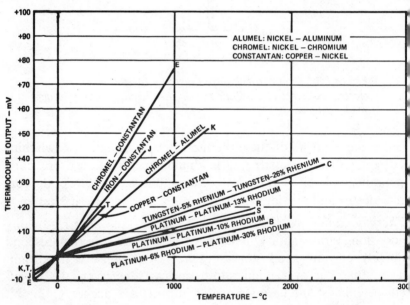

Figure 1-4. Output Characteristics of Thermocouples

Figure 1-4 is a comparative plot of thermocouple output as a function of temperature, referred to a 0°C fixed-temperature reference junction. Figure 1-5a shows a circuit for making a measurement with a thermocouple, using an ice bath to maintain the reference junction at 0°C. Figure 1-5b shows the large number of possible thermocouples in series in a simple circuit.

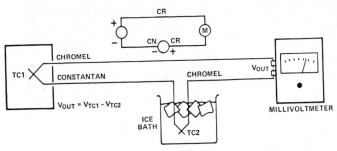

a. Simple temperature measuring circuit using an ice bath at the reference junction. Thermocouple measurements are inherently differential.

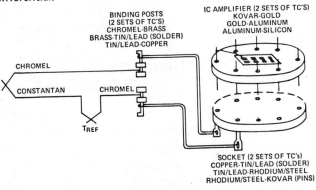

b. The twenty thermocouples in series in a two thermocouple measuring circuit

Figure 1-5. Basic thermocouple circuits

Because thermocouples are low-level (albeit low-impedance) devices, signal conditioning is not a trivial matter. The millivolt-level signals call for low-drift relatively expensive electronics if resolutions better than 1°C are required. Linearity in many types is poor, but the relationships are predictable and repeatable, so either analog or digital techniques can be used for linearizing downstream.

Providing a suitable temperature reference and minimizing the effects of unwanted thermocouples may prove challenging. Techniques include physical references (ice-point cells at +0.01°C, which are accurate and easy to construct but unwieldy to maintain); ambient-temperature reference junctions (acceptable so long as the ambient temperature range in the vicinity of the reference junction is smaller than the desired resolution of the temperature being measured); and electronic cold-junction compensators, which provide an artificial reference level and compensate for ambient temperature variations in the vicinity of the reference junction* (this technique requires careful attention to both the electronics and the physical configuration at that location). Examples of electronic references, which offer good accuracy and require minimal maintenance, will be discussed in Chapter 5 and the Applications section.

The *RTD (Resistance Temperature Detector)* is an electrical circuit element consisting of a solid conductor, usually in the form of wire, characterized by a positive coefficient of resistivity. Platinum, nickel, and nickel-iron alloy are the types in widest use. In general, RTD's are low-level, nonlinear devices with potentially excellent stability and accuracy, when properly implemented and instrumented. Table 1-2 compares several RTD materials.

TABLE 1-2 CHARACTERISTICS OF COMMONLY USED RTD MATERIALS

Material	Temp Range °C	$\approx$T.C.%/°C @ 25°C
Platinum	-200 to +850	0.39
Nickel	-80 to +320	0.67
Copper	-200 to +260	0.38
Nickel-Iron	-200 to +260	0.46

Platinum resistance wire has been generally acknowledged as the standard for accuracy and repeatability in a temperature sensor; it is the standard interpolation device between critical temperatures from -259°C to 631°C. Some units have histories of years of agreement to within a millidegree of reference devices at the calibration facilities at the U.S. National Bureau of Standards. These generic cousins of wirewound resistors are heavily relied upon in transfer-type measurements; though nonlinear, in a predictable way, they can provide linearity to within several degrees over 100°C spans. Operation from -250°C to +850°C is feasible. Units used for standards laboratories can have absolute accuracies to within ±0.001°.

Platinum RTD's are available in ranges from tens of ohms to kilohms with a temperature coefficient of about 0.4%/°C of the resistance value at 25°C. Because they are wirewound, they tend

*In addition to producing a number of modular and system products incorporating cold-junction compensation, Analog Devices expects to have announced a monolithic-IC cold-junction-compensator-preamplifier during 1980.

o be physically large; but platinum-film versions not much larger han 1/8-watt resistors are available at low cost. Platinum sensors or industrial applications have performance approaching that of he standards, but at considerably lower cost. A good industrial ensor can be purchased for much less than $100; a standard with documented history can cost thousands.

Since resistors dissipate energy, platinum sensors require attention to dissipation limitations. Manufacturers provide data on allowable power dissipation for a given accuracy level. The relatively low resistance of platinum sensors requires that the designer consider the potentials used for excitation in voltage dividers or bridges to avoid excessive dissipation. At the same time, the designer must also consider the resistance of the lead wires to the point of measurement. Not only will they have voltage drops; they also have different temperature coefficients. In many cases, current drive, with a separate set of leads carrying no current to measure the voltage directly across the device, is essential.

As noted earlier, platinum is not the only metal used for RTD's, although its high resistance, wide temperature range, and high stability are essential for many applications. Nickel, for example, is popular because it has a relatively high sensitivity (nearly twice that of Pt), is inexpensive, and is usable over a fairly wide range of temperatures. Copper's tempco is similar to that of platinum, but— because of its inherently low resistance—it is difficult to use for precision measurement without extraordinary instrumentation. Other metals, such as gold, silver, and wolfram (tungsten) could be used, in concept, but practical limitations (low resistance for Ag and Au, difficult fabrication for W) prohibit significant usage. RTD's made from a nickel-iron alloy are in use, featuring high resistance, usable tempco (about $0.46\%/°C$), and low cost. Platinum RTD's are now also available in the form of thin film on a ceramic substrate, for reduced size, increased ruggedness, and lower cost.

Thermistors (a contraction for *thermally sensitive resistors*) are electrical circuit elements formed of solid semiconducting materials that are characterized by a high negative* coefficient or resistivity. At any given temperature, a thermistor acts like a resistor; if the temperature changes because of internal dissipation or ambient temperature variations, the resistance changes reproducibly as a

*Though most thermistors have negative temperature coefficients, *positive-temperature-coefficient* (PTC) thermistors are available. Some are characterized by a sharply discontinuous response, useful in on-off control.

function of temperature, in a generally exponential fashion.

They are low in cost and have the highest sensitivity among common temperature transducers (Figure 1-6). At 25°C, a typical unit may have a resistance change of –4.5%/°C. They are manufactured in a range of values from tens of ohms to megohms at 25°C. The response curve is nonlinear but predictable. Typical commercially available sensors provide usable outputs from –100°C to +450°C. Some types can function at temperatures up to 1000°C, with limited accuracy, stability, and sensitivity. Like RTD's, their temperature is manifested electrically by a resistance change, often measured by a bridge circuit. Because of their high sensitivity, they are frequently the best choice in high-resolution measurement-and-control apparatus.

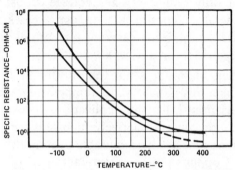

*Figure 1-6. Resistance versus temperature characteristic of two typical thermistor materials (*DESIGN NEWS *11/22/76)*

In the earliest days, thermistors acquired a reputation for being unpredictable and unstable devices, due to problems in manufacture and in application (self-heating effects, for example, could cause wide variations in reading and even thermal runaway). With increased user familiarity and modern manufacturing technologies (which include the production of linearized high-accuracy devices having somewhat less sensitivity and higher cost), they can be used with a great deal of confidence.

For example, the Yellow Springs YSI 44000-series sensors have interchangeability and uniformity to within 0.1°C from –40°C to +100°C and 0.2% from –80°C to +150°C, with cost less than $12*. Thermometrics makes devices that have accuracy and long-term-stability specifications comparable with those of platinum

*At the time this was written. Because prices vary, readers should consider any prices mentioned here as indicative of relative (rather than absolute) cost.

(at comparable cost) but with greater sensitivity, thus simplifying the signal conditioning. Absolute accuracy can run to within 0.01°C from 0°C to 60°C.

Thermistors are generally quite small and fast response is a typical feature. A common *bead* unit may have a time constant measured in seconds, while small *flakes*, such as those manufactured by Veeco, can respond in milliseconds.

Linearized thermistors have two or more devices in a single package, used with fixed resistors, to provide output potentiometrically (3-terminal network) or as a linear resistance variation, as Figure 1-7 shows. Thermistors are well-suited to sensitive bridge-type measurements because of their high (and selectable) impedance. As with RTD's, thermistors have dissipation limitations which must be observed if desired optimum performance is to be obtained, since internal heating affects accuracy. On the other hand, self-heating is used in feedback null-type linearity-independent measurements.

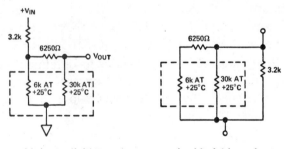

a. *Voltage divider*   b. *Variable resistor*

Figure 1-7. Linearized thermistors. Fixed resistors and two thermistors are arranged in a network (Yellow Springs Inst. CO 44000 series) to provide linear, accurate response. In "A" the network has three terminals and functions as a temperature dependent voltage divider. In "B" the same network is used in the two-terminal mode to provide a linear shift in resistance vs. temperature.

Semiconductor Sensors, generally based on the temperature sensitivity of silicon devices, are economical and available in many forms.* We will limit the discussion here to devices having two terminals, since they are usually the most suitable devices for remote measurement. The three classes of device to be mentioned

*Some circuit designers are convinced that there are too many incidental forms!

are bulk resistors (e.g., "Tempsistors"), diodes, and integrated circuits (e.g., the AD590).

The simplest form of semiconductor temperature sensor consists of a piece of bulk silicon. Devices of this type are available at low cost. They feature a positive temperature coefficient, about 0.7%/°C, and linearity to within ±0.5% from –65°C to 200°C. Nominal resistance ranges from 10Ω to 10kΩ, with tolerance from 1% to 20%. Physically, they look like 1/4-watt resistors. Since operation is specified at zero power (i.e., with no current flowing through the device), self-heating effects must be taken into account. Like other resistive devices, silicon resistors may be used in bridge circuits.

Junction semiconductor devices are well-suited to temperature measurement. The junction potential of silicon transistors and diodes, though differing from device to device, changes at about 2.2mV/°C over a wide range of temperature and can be used as the basis of an inexpensive sensor having fast response. Since diode voltage is also a function of current, the source of excitation may be a constant current. Figure 1-8 shows the relationships between temperature and V_{BE} for Motorola MTS105-series devices having several initial values of V_{BE}. In order to obtain accurate output, the diodes must be either calibrated or used in matched pairs in bridge-type circuits; though diodes are low in cost, these considerations make them less competitive.

(1) $TC = -2.25 + 0.0033 (V_{BE} - 600)$

(2) $V_{BE(T_A)} = V_{BE(25°C)} + (TC)(T_A - 25°C)$

1. Determine Value of TC:

2. Determine the V_{BE} Value at the Extremes, –40°C and +150°C:

3. Plot the V_{BE} Versus T_A Curve Using Two V_{BE} Values: $V_{BE(-40°C)}$, $V_{BE(25°C)}$, or $V_{BE(+150°C)}$.

4. Given Any Measured V_{BE}, the Value of T_A Can Be Read from the Above Curve, or Calculated from Equation 2.

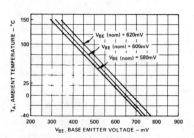

Figure 1-8. Using a diode as a temperature sensor: Motorola MTS105 series

Two-terminal temperature-sensitive current sources in the form of monolithic IC's are exemplified by the Analog Devices AD590.

Available in cans, miniature flat packages, chip form, and stainless steel probes, it is a current source which passes a current numerically equal (microamperes) to absolute temperature (kelvin), when excited by a voltage from +4V to +30V (Figure 1-9a), at temperatures from −55°C to +150°C. Figure 1-9b is a simplified schematic, which shows how it works. Figure 1-9c shows how it might be simply used to implement a remote measurement.

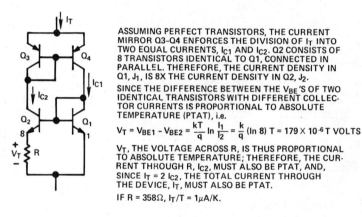

$I(\mu A) = T(K)$ 4V ≤ V ≤ 30V

−55°C ≤ T ≤ +150°C
−67°F ≤ T ≤ +302°C
218.2K ≤ T ≤ +423.2K
218.2μA ≤ I ≤ 423.2μA

a. AD590 as a 2-terminal device

ASSUMING PERFECT TRANSISTORS, THE CURRENT MIRROR Q3–Q4 ENFORCES THE DIVISION OF I_T INTO TWO EQUAL CURRENTS, I_{C1} AND I_{C2}. Q2 CONSISTS OF 8 TRANSISTORS IDENTICAL TO Q1, CONNECTED IN PARALLEL. THEREFORE, THE CURRENT DENSITY IN Q1, J_1, IS 8X THE CURRENT DENSITY IN Q2, J_2.

SINCE THE DIFFERENCE BETWEEN THE V_{BE}'S OF TWO IDENTICAL TRANSISTORS WITH DIFFERENT COLLECTOR CURRENTS IS PROPORTIONAL TO ABSOLUTE TEMPERATURE (PTAT), i.e.

$$V_T = V_{BE1} - V_{BE2} = \frac{kT}{q} \ln \frac{I_1}{I_2} = \frac{k}{q} (\ln 8)\, T = 179 \times 10^{-6}\, T \text{ VOLTS}$$

V_T, THE VOLTAGE ACROSS R, IS THUS PROPORTIONAL TO ABSOLUTE TEMPERATURE; THEREFORE, THE CURRENT THROUGH R, I_{C2}, MUST ALSO BE PTAT, AND, SINCE $I_T = 2\,I_{C2}$, THE TOTAL CURRENT THROUGH THE DEVICE, I_T, MUST ALSO BE PTAT.

IF R = 358Ω, I_T/T = 1μA/K.

b. How the AD590 works—simplified circuit

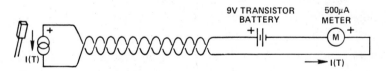

c. Simple implementation of the AD590

Figure 1-9. Absolute temperature-to-current IC transducer

The AD590 has a standardized (1μA/K) output (in several accuracy grades—see Appendix); it embodies an inherently linear relationship and is easy to use, not requiring bridges, low-level voltage measurement, or linearizing circuitry. Since its output is a current, long leads may be used without errors due to voltage drops or induced voltage noise; and since it is a high-impedance current source, it is insensitive to excitation voltage (in order to minimize self-heating effects, use the lowest excitation voltage consistent with the desired output voltage and expected line drops; however, even in the worst case, the maximum dissipation is only 13mW).

It may be worth noting that digital meters (AD2040) and scanners (AD2038), for directly instrumenting measurements in kelvin, °F, and °C, are available from Analog Devices. Measurements in °C are implemented by subtracting (in effect) a fixed current of 273.2μA; measurements in °F require subtraction of 255.4μA* and scaling by 9/5. Differential Celsius measurements with reference to any fixed temperature are obtained by subtracting a current numerically equal to the absolute (kelvin) equivalent of that temperature.

Force Transducers

The most popular electrical elements used in force measurements include the resistance strain gage, the semiconductor strain gage, and piezoelectric transducers; they are described briefly below. In general, the strain gage measures force indirectly by measuring the deflection it produces in a calibrated carrier; the piezoelectric transducer responds directly to the force applied. Implementation in bridges will be described in the next chapter; and further details of their application will be found in the descriptions in the Applications section.

The *resistance strain gage* is a resistive element which changes in length, hence resistance, as the force applied to the base on which it is mounted causes stretching or compression. It is perhaps the most well-known transducer for converting force into an electrical variable.

Unbonded strain gages consist of a wire stretched between two points. Force acting on the wire will cause the wire to elongate or shorten, which will cause the resistance to increase or decrease ($R = \rho L/A$, $\Delta R/R = K \Delta L/L$, where K is the gage factor†).

Bonded strain gages consist of a thin wire of conducting film arranged in a coplanar pattern and cemented to a base or carrier.

---

*°C = K − 273.2, and °F = $\frac{9}{5}$ °C + 32°. Hence,

°F = $\frac{9}{5}$ (K −273.2 + $\frac{5}{9}$ ∘ 32°) = $\frac{9}{5}$ (K − 255.4°)

†Gage factor is a function of the conductor material, ranging from a minimum of 2.0 to 4.5 for metals and more than 150 for semiconductors.

The gage is normally mounted so that as much as possible of the length of the conductor is aligned in the direction of the stress that is being measured. Lead wires are attached to the base and brought out for interconnection. Bonded devices are considerably more practical and are in much wider use than the unbonded devices. Figure 1-10 shows several typical patterns in use.

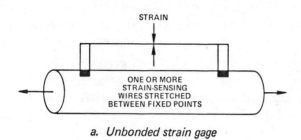

a. Unbonded strain gage

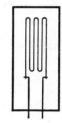

b. Bonded wire strain gage

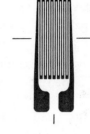

c. Foil strain gage

d. 2-Element rosette 90°
planar foil strain gage

e. 3-Element rosette 60°
planar foil strain gage

Figure 1-10. Strain gages. Multiple-element rosettes measure components of strain in different directions. For example, the elements at 90° can measure magnitude and direction of stretch. Many patterns with various numbers and configurations of elements are available. (Courtesy of BLH Electronics—SR-4 Strain Gage Handbook)

A great deal of effort has been devoted to making the strain gage the reliable and stable device it is today. Though the relationship between mechanical strain and electrical resistance change was observed by Lord Kelvin in 1856, more than 75 years elapsed before the effect was utilized in high-performance devices. Perhaps the most popular modern adaptation is the foil-type gage, produced by photo-etching techniques, and using similar metals to the wire types (alloys of copper-nickel (Constantan), nickel-chromium (Nichrome), nickel-iron, platinum-tungsten, etc.).

Gages having wire sensing elements present a small surface area to the specimen; this reduces leakage currents at high temperatures and permits higher isolation potentials between the sensing element and the specimen. Foil sensing elements, on the other hand, have a large ratio of surface area to cross-sectional area; they are more stable under extremes of temperature and prolonged loading. The large surface area and thin cross section also permit the device to follow the specimen temperature and facilitate the dissipation of self-induced heat.

The *load cell* is a commonly used form of strain-gage-based transducer. It converts an applied force (weight) into a bridge output potential. In a load cell, the strain gage is mounted on some form of mechanical sensing element (column, beam, etc.), and the gage (or gages) is (are) usually wired into a bridge configuration. Compensation for temperature and nonlinearity is provided for by the manufacturer in the selection of resistance values for the arms of the bridge and in series with the bridge.

Strain gages are low-impedance devices; they require significant excitation power to get output voltage at reasonable levels. A typical strain-gage-based load cell will have a 350Ω impedance and is specified as having a sensitivity in terms of *millivolts per volt of excitation at full scale*. The maximum excitation potential, as well as the recommended potential, will be specified. For a 10V device with a rating of 3mV/V, 30 millivolts of signal will be available at full-scale loading. The output can be increased by increasing the drive to the bridge, but self-heating effects are a significant limitation to this approach: they can cause erroneous readings or even device destruction if prolonged.

The low output of load-cell transducers in most cases is due to the small shifts of resistance in the strain gages. However, while this is a severe constraint on gain, it is a boon to linearity (as will be shown, the output of a bridge is linear only for small changes in resistance, in the most-frequently used configurations, even if the gage resistance-change is perfectly linear with applied force).

The low impedance employed in strain-gage bridges may require remote sensing for fixed excitation. Because the bridge operates off-null, shifts or inaccuracies in the excitation voltage will contribute directly to shifts in the bridge output. The low impedance of the bridge means that voltage drops in the wires leading to the bridge input can contribute significantly to variations in excita-

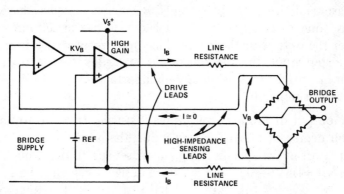

a. Basic scheme. Voltage across bridge terminals is compared with reference; high-gain feedback loop causes output of bridge supply to be at whatever voltage is necessary for null ($V_{REF} - KV_B = 0$) at comparator input, hence $V_B = V_{REF}/K$

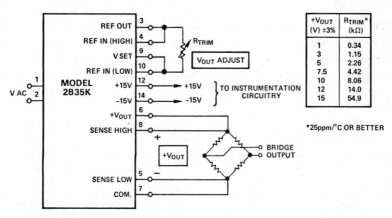

+V$_{OUT}$ (V) ±3%	R$_{TRIM}$* (kΩ)
1	0.34
3	1.15
5	2.26
7.5	4.42
10	8.06
12	14.0
15	54.9

*25ppm/°C OR BETTER

b. Regulated excitation voltage, from +1V to +15V, using the 2B35 transducer supply

Figure 1-11. Remote bridge drive using voltage sensing

tion, hence error. To correct this, manufacturers often employ four-wire (Kelvin) connections to the bridge input. Two wires carry the bridge current, and two wires sense the actual voltage at the bridge; the fed-back voltage is compared with the reference, and the power-supply output is adjusted to whatever voltage is necessary to maintain the bridge voltage at the desired value (Figure 1-11).*

Load cells are manufactured in capacities ranging from pounds to kilotons (kg to Gg). Response time is usually limited to periods measureable in tenths of seconds. Responses also may have "long tails" due to *creep*, a small, persistent change in cell output well after the output of the load cell has apparently settled in response to a step input; the phenomenon is analogous to dielectric absorption in capacitors. It is usually unimportant, except for high-precision measurements.

The output of the load-cell bridge is usually trimmed to furnish a small degree of offset for zero input; the magnitude and polarity of the offset are specified (e.g., an offset specification might read: "Offset +1mV − 0". The interface designer can then be sure that all load cells of this type will have offsets of the same polarity, a fact that simplifies the zero-adjustment in the interface. Linearity, temperature coefficients, overrange capability,† and other self-explanatory specifications are usually provided. Typical manufacturers of load cells are BLH Electronics, NCI, and Transducers, Inc.

Semiconductor strain gages make use of the resistance change in semiconductor materials in order to obtain greater sensitivity and higher-level output. Such bridges may have 30 times the sensitivity of bridges employing metal films, but they are temperature-sensitive and are not easy to compensate. They have not come

---

*The 2B31 signal conditioner and the 2B35 transducer power supply both utilize sense feedback for precision voltage or current excitation.

†It is common to specify the maximum static load which will not damage the device. However, it is worth noting that transient loading must also be considered. A 150-pound-maximum-load transducer, which will be damaged by a 300-lb static load might also be damaged by a 75-lb load dropped it from a distance of one or two feet, since the instantaneous force developed by rapid deceleration from the speed reached due to gravitational pull may greatly exceed the specification. This is not exactly an academic afterthought: if one considers the instantaneous force generated by dropping a large pumpkin on an inadequately damped supermarket electronic scale, one can appreciate the plight of a scale manufacturer (perhaps legendary) who suffered a flurry of warranty replacements after a disastrous Halloween season. Cures for such situations might include more-adequately specified load cells, increased damping, or mechanical stops.

into as widespread use as the more stable metal-film devices for precision work; however, where sensitivity is important and temperature variations are not great, and in null-type measurements where the bridge is always in balance, they may have some advantage. Instrumentation is similar to that for metal-film bridges but is less critical because of the higher signal levels and decreased transducer accuracy.

Piezoelectric force transducers are employed where the forces to be measured are dynamic (i.e., continually changing over the period of interest—usually of the order of milliseconds). These devices utilize the effect discovered by Pierre & Jacques Curie in 1880, that changes in charge are produced in certain materials when they are subjected to physical stress. Piezoelectric devices produce substantial output voltage in instruments such as accelerometers for vibration studies. Output impedance is high, and charge amplifier configurations, with low input capacitance, are required for signal conditioning.

The output of a piezoelectric transducer may be modeled as a voltage source in series with a small capacitor. Step inputs of physical force result in an effective capacitance change. When instrumented with feedback amplifiers (op amps), as in Figure 1-12, the summing-point voltage is held at zero, and the change in charge is effectively transferred to the feedback capacitor, developing an output voltage at low impedance. The output of this circuit is inversely proportional to the value of feedback capacitance. Practical examples and considerations in the design of the charge amplifier appear in the Applications section. Manufacturers of piezoelectric devices often furnish calibrated charge amplifiers, cables, and other accessories.

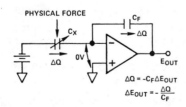

Figure 1-12. Charge Amplifier—commonly used in piezoelectric accelerometers

Piezoelectric force transducers are available from Endevco, Columbia Research Labs, Kistler Instrument Corporation, and others.

Pressure Transducers

Pressures in liquids and gases are measured electrically by a variety of pressure transducers. A variety of mechanical converters, including diaphragms, capsules, bellows, manometer tubes, and Bourdon tubes, are used to measure pressure by measuring an associated length, distance, displacement, and to measure pressure *changes* by the motion produced. As with most transducers, a great deal of expertise (some would say, aided by witchcraft) is required to obtain a stable, accurate, linearly responding pressure transducer. The output of this mechanical interface is then applied to an electrical converter.

Rheostats and potentiometers are often used to convert linear or rotary motion to an electrical output. A rheostat, like a strain gage, simply produces a varying resistance over a range of pressure inputs. A potentiometer may be used as a pair of series arms in a linear-output bridge configuration. Temperature effects are minimized by the use of wirewound rheostats, but resolution may suffer as the wiper arm indexes between adjacent turns of multi-turn wirewound elements. Conductive plastic and metal films achieve higher resolution at some cost in stability with temperature. Potentiometric elements provide good tracking, and temperature is not troublesome.

Strain gages are also used in pressure transducers. The mechanical output of the transducer produces a resistance change in a strain gage, which is configured electrically in similar manner to a load cell.

Piezoelectric pressure transducers are used for high-frequency pressure measurements. They are also employed in sound-level pressure measurement (and are better-known here as *crystal microphones*). Signal conditioning for piezoelectrics involves high-impedance (voltage mode) or charge-type amplifiers.

There are three general categories of pressure measurement—absolute, gauge, and differential (Figure 1-13). *Absolute*-pressure devices measure pressure with reference to zero pressure (i.e., vacuum). *Gauge* pressure is measured in relation to ambient (which may be standard sea-level atmospheric pressure, or an arbitrary level). *Differential*-pressure transducers measure the difference between two pressures (for example, across a valve or an orifice). In a sense, a gauge measurement is a special form of differential measurement in which one of the pressures, being ambient, does not require special provisions for connections.

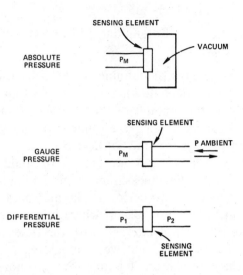

Figure 1-13. Three types of pressure transducer

Specifications regarding accuracy, temperature coefficient, linearity, etc., are usually self-explanatory or are defined by the manufacturer. The range of pressures may be unipolar ("pressure" or "vacuum") or bipolar (positive or negative), and the measurement may be offset to cover a limited range, for example 75-to-85psi. Piezoelectric types respond most readily to fast changes. In some applications, response to rapid changes is undesirable, and damping is provided ("noise filtering"). In many applications, changes are slow and considerations such as frequency or time response are irrelevant.

One of the most frequent sources of error in the application of pressure transducers is inappropriate application within a given medium. Many transducers are intended only for use with gases and liquids which are non-corrosive or in other ways benign to the transducer. The system designer should always determine from the manufacturer what substances may come into contact with the transducer (it may be necessary to place a buffer substance between the measurand and the transducer—a more common necessity than one might suspect, and one that is easily implemented.

Another frequently overlooked consideration is the temperature of the measurand. While the ambient may be at 30°C, the inlet temperature of the liquid or gas at the transducer might be 125°C, in which case transducer selection or rebudgeting of allowable error tolerances may be required.

Manufacturers include Rosemount, BLH, and others.

Flow Transducers

There are many ways of defining flow (mass flow, volume flow, laminar flow, turbulent flow). Since the "bottom line" of a flow measurement is the amount of the substance flowing that is useful for some purpose (e.g., the number of molecules of hydrocarbon available for combustion), the desirable measurement is most-often *mass flow* (kg/s); however, if the fluid's density varies but little, a *volume flow* measurement (m³/s) is a useful substitute that is generally easier to perform. The measured flow may vary, depending on the type of sensor, where it is located in the stream and the way in which it interacts with the fluid. There are many ways of measuring flow, and we cannot even pretend to do lip service to them. However, it is important to note that electrical outputs in the form of voltage, current, or frequency tend to be handled in much the same way, irrespective of the physical mechanism involved in the measurement.

One commonly used class of transducers, which measure flow rate indirectly, involves the measurement of pressure. Flow can be derived by taking the differential pressure across two points in a flowing medium—one at a static point and one in the flow stream. *Pitot tubes* (Figure 1-14) are one form of device used to perform this function. The flow rate is obtained by measuring the differential pressure with standard pressure transducers, and calibrating (or otherwise dealing with) the nonlinear relationship.

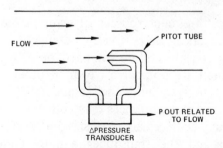

Figure 1-14. Pitot tube/static pressure flow measurement

Some transducers produce frequency signals. Examples include propellers, turbines, "positive displacement" meters, for accurate measurement of quantity, and various paddle-wheel arrangements, as well as vortex-shedding obstructions. The frequency signals, picked off electrically, optically, or magnetically, can be directly transduced to digital form.

The cantilevered vane (Figure 1-15A) is simple and amenable to strain-gage instrumentation. The hinged vane is also simple and works well with potentiometers (Figure 1-15b).

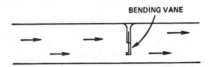

Figure 1-15a. Cantilevered vane flow meter

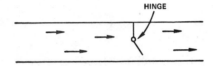

Figure 1-15b. Hinged vane flow meter

The application considerations for flow transducers using pressure transducers are the same as for the latter. Among frequency-output types, some require signal conditioning, while others have direct high-level pulses available at the output. Very low flow-rate detection with frequency-output devices is sometimes difficult or tedious because of the low resolution at low frequencies.

Anemometers comprise a special class of flowmeters, which are used almost exclusively to measure wind speed. Anemometers using propellors or cups usually drive some sort of tachometric device which interfaces to the readout. *Hot-wire* anemometers (and flowmeters) consist of a heated wire, supported at the ends, which loses heat to the fluid stream being measured. This convective loss varies approximately with the square-root of velocity. The resistance of the heated wire is measured and used to provide a readout. In another form of circuitry, a feedback circuit is employed to maintain the wire at constant temperature through self-heating; the power input to the wire is then a good measure of the wind speed, and response to changes in speed can be quite fast since only the electrical input power changes, not the temperature.[1]

---

[1] *Analog Dialogue*, Volume 5, No. 1, page 13, "Measuring Air Flow Using a Self-Balancing Bridge," by José Miyara

Hot-wire techniques have also been used to measure speed of ships through water.

Other types of flowmeters include electromagnetic and ultrasonic Doppler, especially for non-invasive measurements and for supersonic flow. They are generally supplied as complete instruments.

Level Transducers

A better name for these devices is "volume transducers", since level transducers are most-often used to measure the contents of containers. The best-known level transducer (except for the calibrated stick) is the float type, using in millions of automobile gas tanks. A float controls a potentiometer or a rheostat, which provides an electrical output. The output may be discrete as well as continuous, if the potentiometer is replaced by a set of switch contacts.

The liquid itself may be used as the "rheostat" if the conductance between two rods, immersed in the liquid, is measured (Figure 1-16a). Discrete level information may be taken similarly, as Figure 1-16b shows. Capacitance may also be used as the parameter. Measurements are ac-type (easy to amplify), and the method will work with conductive or nonconductive materials—wet or dry.

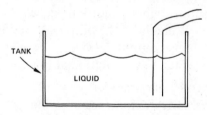

Figure 1-16a. Sensing level by sensing conductance or capacitance

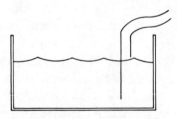

Figure 1-16b. Discrete level sensing

A popular method for discrete level detection involves the principle that the rate of heat transfer is much greater in a liquid than in a gas. Thermistors and other temperature detectors, if deliberately operated in a self-heating mode when surrounded by a gas, will show a pronounced shift in temperature reading when in contact with a liquid. This shift can be detected by simple electronic circuitry which produces contact closures or any desired form of indication.

Discrete sensing can be performed optically by detection of the state of an optical path (Figure 1-17). The presence of fluid causes scattering or absorption of light, which breaks up the optical path.

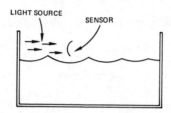

Figure 1-17. Discrete level sensing by optical scattering

Pressure transducers can be used to determine the level in a tank by measuring the differential pressure between the unoccupied area in the top of the tank and the liquid-covered area (Figure 1-18). The level will be directly proportional to the differential pressure for any given specific weight of liquid in the cylindrical tank. The electrical considerations for pressure transducers were described earlier in this chapter.

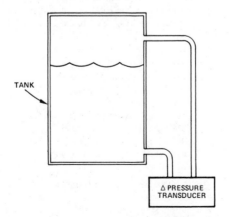

Figure 1-18. Level-by-differential pressure

In many systems, level (or *mass*) is sensed accurately by simply weighing the tank and subtracting the necessary offsets (*tare* in weighing terminology). Load-cell transducers are almost always used in such applications.

Acoustic (sonar) techniques, involving an echo-delay measurement are in use. The method works well with both liquids and solids, in both discrete and continuous versions. Another technique (among many) is a floating ring magnet surrounding a protective insulating vertical tube containing a string of magnetically operated reed relays; a thermal version of this utilizes a string of AD590 temperature sensors.

COMMON TRANSDUCERS SUMMARIZED

TYPE	TEMPERATURE ELECTRICAL I/O CHARACTERISTICS	COMMENTS
Thermoswitches	Switch closure. Simple on-off output	Many types available, covering a wide range of temperatures, contact configurations, and current-handling capabilities.
Thermocouples	Low source impedance, typically 10Ω. Voltage-output devices. Output shift is 10's of microvolts/$^\circ$C. Outputs typically in the millivolts at room temperature.	Low voltage output requires low-drift signal conditioning. Small size and wide temperature range are advantages. Requires reference to a known temperature. Nonlinear response.
Platinum and other RTD's	Resistance changes with temperature. Positive temperature coefficient. Typical impedance (0°C) 20Ω to 2kΩ. Typical sensitivities 0.1%/$^\circ$C to 0.66%/$^\circ$C, depending on material.	Highly repeatable. Good linearity over wide ranges. Requires bridge or other network for typical interface.
Thermistors	Resistance changes with temperature. Negative temperature coefficient. Typical impedances (25°C) of 50Ω to 1MΩ available. Sensitivity at 25°C is about 4%/$^\circ$C. Linearized networks available with with 0.4%/$^\circ$C sensitivity.	Highest sensitivity among common temperature transducers. Inherently nonlinear (exponential function) but accurate linearized networks available.
Semiconductor sensors	Voltage, current, or resistance functions. Voltage types (diodes) require excitation current. Current types (AD590) require excitation voltage. Resistive types (bulk silicon) may use either type of excitation.	Many devices are uncalibrated and require significant signal conditioning. AD590 is calibrated, linear, and requires minimal signal conditioning.

FORCE

TYPE	ELECTRICAL I/O CHARACTERISTICS	COMMENTS
Strain gages (metal)	Resistance shifts with applied strain. Almost always used in bridge configuration. Typical impedance levels of 120Ω and 350Ω. Typical change is 0.1% over the whole range.	Resistance change with strain small compared to initial value of device resistance. Requires high-quality low-level signal conditioning.
Strain-gage bridge, load cell	Voltage output with applied strain. Requires excitation potential or current to drive the bridge. Typical excitation is from 5 to 15 volts.	Small voltage outputs require low-drift signal conditioning with good common-mode rejection to achieve any degree degree of precision. Output is linear.
Semiconductor strain gages	Bridge types are assembled from individual gages and have a voltage output. Bridge requires excitation, typically 5V to 15V.	More output than metal strain gages, but with increased non-linearity and sensitivity to temperature.
Piezoelectrics	True charge output device. Modeled as voltage source in series with capacitor. Physical input change produces corresponding charge change. AC and transient response only. Typical upper frequency limit is 20 to 50kHz. Typical output is 10^{-7} coulombs full-scale.	Requires low-bias-current charge amplifier configurations for signal conditioning. Responds to ac signals only.

PRESSURE

TYPE	ELECTRICAL I/O CHARACTERISTICS	COMMENTS
Rheostat/potentiometer	Resistance or ratio-of-resistance output. Requires voltage or current excitation. Typical impedance 500Ω to 5kΩ.	High-level easy-to-condition outputs are typical due to significant resistance or ratio
Strain gage	Resistance shift (single gage) or voltage output (strain-gage bridge). Requires excitation potential or current.	Small resistance change. Low-level signal requires good signal-conditioning amplifiers.
Piezoelectric	Charge output (see FORCE transducer chart).	See FORCE

FLOW

TYPE	ELECTRICAL I/O CHARACTERISTICS	COMMENTS
Pressure-based	See PRESSURE transducers	Pressure types measure flow by measuring ΔP between static and flow-caused pressure, or pressure drop across a constriction. Differential pressure transducers are used to avoid common-mode pressure errors. Response is nonlinear.
Frequency-output types: paddle wheels, rotary types, vortex types	Digital output derived from frequency output are common. Optical or magnetic pickups provide non-invasive measurements. Photocell has 100Ω to $100M\Omega$ on-to-off ratio. Magnetic employs switching or open-collector transistor.	Some types are directly logic-level compatible. Others require impedance and/or voltage amplification, level-shift, and buffering before signal is usable.
Force-based	Typical forms use strain-gage bridges or potentiometer outputs. See PRESSURE and FORCE.	See PRESSURE and FORCE
Thermal	Use active temperature sensors to measure temperature changes caused by flow	See TEMPERATURE

LEVEL

TYPE	ELECTRICAL I/O CHARACTERISTICS	COMMENTS
Float	Resistor or potentiometer output. 100Ω to $2k\Omega$ typical impedance.	Requires excitation (current, voltage) to achieve voltage output. High-level output due to large resistance swings.
Thermal	Resistive. Typical impedances 500Ω to $2k\Omega$.	Self-heated temperature sensor (thermistor) is used to detect discrete level changes. Abrupt resistance changes occur when liquid level drops to allow thermistor to be uncovered.
Optical	Resistive. Typical on-off impedances 100Ω to $100M\Omega$.	Optical occlusion or scattering blocks an opto-electronic path.
Pressure	See PRESSURE	Level information obtained by measuring pressure in un-occupied area in top enclosed tank vs. pressure in liquid-covered area.
Load cell	Contents of container measured by weighing	See FORCE

Interfacing Considerations
Bridges

Chapter 2

Figure 2-1 shows the common Wheatstone bridge (actually developed by S. H. Christie in 1833). In its simplest form, a bridge consists of four two-terminal elements connected to form a quadrilateral, a source of excitation (voltage or current)—connected along one of the diagonals, and a detector of voltage or current—comprising the other diagonal. The detector, in effect, measures the difference between the outputs of two potentiometric dividers connected across the excitation supply.

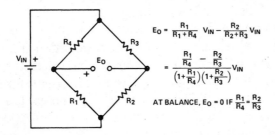

$$E_O = \frac{R_1}{R_1+R_4}\, V_{IN} - \frac{R_2}{R_2+R_3}\, V_{IN}$$

$$= \frac{\dfrac{R_1}{R_4} - \dfrac{R_2}{R_3}}{\left(1+\dfrac{R_1}{R_4}\right)\left(1+\dfrac{R_2}{R_3}\right)} V_{IN}$$

AT BALANCE, $E_O = 0$ IF $\dfrac{R_1}{R_4} = \dfrac{R_2}{R_3}$

Figure 2-1. Basic bridge circuit — voltage excitation and voltage readout

A bridge measures an electrical property of a circuit element indirectly, i.e., by comparison against a similar element. The two principal ways of operating a bridge are as a null detector and as a device that reads a difference directly in voltage or current.

When $R_1/R_4 = R_2/R_3$, the resistance bridge shown in Figure 1 is at a *null*, irrespective of the mode of excitation (current or voltage, ac or dc), the magnitude of excitation, the mode of readout (current or voltage), or the impedance of the detector. Therefore, if the ratio R_2/R_3 is fixed at K, a null is achieved when $R_1 = K R_4$. If R_1 is unknown and R_4 is an accurately determined variable resistance, the magnitude of R_1 can be found by adjusting R_4 until null is achieved. Conversely, in transducer-type measurements, R_4 may be a fixed reference and a null occurs when the magnitude of the measurand is such that R_1 is equal to $K R_4$.

Null-type measurements are principally used in feedback systems, involving electromechanical and/or human elements. Such systems, as noted in the previous chapter, seek to force the active element (strain gage, RTD, thermistor, mechanically coupled potentiometer) to balance the bridge by influencing the parameter being measured. Because the null is independent of the excitation, the null mode may also be used to discriminate between the two polarities of output, i.e., as a *comparator*. In such applications, the *polarity* of the off-null signal might be of greater significance than its *magnitude* (for example, if the level of a tank is below a preset value, a valve is caused to open to fill the tank).

For the majority of transducer applications employing bridges, the *deviation* of one or more resistors in a bridge from an initial value must be measured as an indication of the magnitude (or a change) of the measurand. Figure 2-2 shows a bridge with all resistances nominally equal; but one of them (R_1) is variable by a factor, $(1 + X)$, where X is a fractional deviation around zero, as a function of (say) strain. As the equation indicates, the relationship between the bridge output and X is not linear, but for small ranges of X it is sufficiently linear for many purposes. For example, if $V_{IN} = 10V$, and the maximum value of X is ±0.002, the output of the bridge will be linear to within 0.1% for a range of outputs

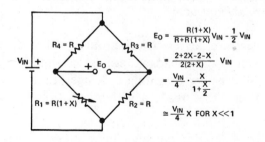

$$E_O = \frac{R(1+X)}{R+R(1+X)} V_{IN} - \frac{1}{2} V_{IN}$$

$$= \frac{2+2X-2-X}{2(2+X)} V_{IN}$$

$$= \frac{V_{IN}}{4} \cdot \frac{X}{1+\frac{X}{2}}$$

$$\cong \frac{V_{IN}}{4} X \text{ FOR } X \ll 1$$

Figure 2-2. Bridge used to read deviation of a single variable element

from 0 to ±5mV, and to 1% for the range 0 to ±50mV (±0.02 range for X).

The *sensitivity* of a bridge is the ratio-to-the-excitation-voltage of the maximum expected change in the value of the output; in the examples given in the last paragraph, the sensitivities are ±500μV/V and ±5mV/V. The sensitivity can be doubled if two identical variable elements can be used, e.g., at positions R3 and R1, as shown in Figure 2-3a. An example of such a pair is two identically oriented strain-gage resistances aligned in a single pattern. Note that the output is doubled, but the same degree of non-linearity exists.

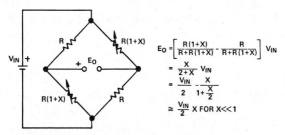

$$E_O = \left[\frac{R(1+X)}{R+R(1+X)} - \frac{R}{R+R(1+X)} \right] V_{IN}$$

$$= \frac{X}{2+X} V_{IN}$$

$$= \frac{V_{IN}}{2} \cdot \frac{X}{1+\frac{X}{2}}$$

$$\cong \frac{V_{IN}}{2} X \quad \text{FOR } X \ll 1$$

Figure 2-3a. Bridge with two variable elements

In special cases, another doubling of the output can be achieved. Figure 2-3b shows a bridge consisting of four resistors, two of which increases and two of which decrease in the same ratio. Two identical two-element strain gages, attched to opposite faces of a thin carrier to measure its bending, could be electrically configured in this way. The output of such a bridge would be four times the output for a single-element bridge; furthermore, the complementary nature of the resistance changes would result in a *linear* output.

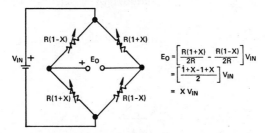

$$E_O = \left[\frac{R(1+X)}{2R} - \frac{R(1-X)}{2R} \right] V_{IN}$$

$$= \left[\frac{1+X-1+X}{2} \right] V_{IN}$$

$$= X V_{IN}$$

Figure 2-3b. All elements variable

Figure 2-3c shows a bridge employing a zero-centred potentiometer to constitute two adjacent arms; the position of the potentiometer is a measure of the physical phenomenon. Since it is a 2-variable-element version of 2-3b, the output is twice that of the single-element bridge, and it is linear.

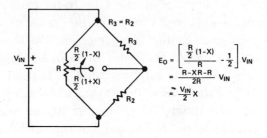

c. Linear potentiometer as variable arm

Figure 2-3. Useful bridge configurations

A distinction should be recognized between the linearity of the bridge equation and linearity of the transducer response to the phenomenon being sensed. For example, if the active element is a potentiometer, a bridge used to implement the measurement would be adequately linear; yet the output could still be nonlinear due to the pot's nonlinearity.

Manufacturers of transducers employing bridges address the non-linearity issue in a variety of ways, including keeping the resistive swings in the bridge small, shaping complementary nonlinear response into the active elements of the bridge, using resistive trims for first-order corrections, and a variety of proprietary magical techniques. A bridge can, of course, be linearized by making it less sensitive (e.g., by making the initial ratios, R_4/R_1 and R_3/R_2, large), but the tradeoff of sensitivity for linearity is painful.

Figure 2-4 shows an active bridge in which an op amp produces

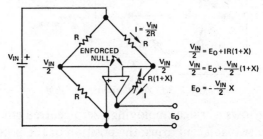

Figure 2-4. Active bridge

a null by adding a voltage in series with the variable arm. That voltage is equal in magnitude and opposite in polarity to the incremental voltage across R_X, and it is inherently linear with X. Since it is an op-amp output, it can be used as a low-impedance output point for the bridge measurement. This active bridge has a gain of two over the standard one-active-element bridge, and the output is linear, even for very large values of X.

More information about linearization techniques can be found in Chapter 5.

EXCITATION

The choice of circuitry to produce the excitation voltage (or current) will depend on the system designer's background and inclinations and will depend on the degree of precision and any special requirements for the specific system. Details of a variety of approaches are shown in the Applications section. Choices range from assemblages of components to "system solutions."*

A stable bridge-driving potential may be obtained through the use of reference IC's and op-amp circuitry (Figure 2-5a). The particularities of op-amp circuitry can be avoided by the use of a fully engineered and specified signal-conditioning power supply, which provides resistor-programmable current or voltage and has sensing leads that permit precise voltage to be maintained at the bridge terminals in spite of voltage drops in the leads (Figure 2-5b). For minimum component count, complete signal conditioners provide programmable excitation, in addition to amplification and filtering.*

For high-precision measurements, the requirement for a highly stable and accurate supply may be less pressing if the same reference can be used for both the bridge and the readout device (e.g., a digital panel meter). Such measurements, in which the *ratio to full scale* for both devices is accurately maintained, are independent of the actual level of the excitation; not surprisingly, this technique is known as *ratiometric* measurement (Figure 2-5c).

READOUT

The hardware to detect and measure the output from a bridge can

---

*It will be difficult to maintain the appearance of objectivity in the discussion of hardware, since it is available in many guises from Analog Devices. We will seek to avoid references to Model numbers in the text, unless absolutely necessary. However, since they constitute useful real-world information, they will be used liberally in circuit diagrams and, where appropriate, in footnotes, throughout the book.

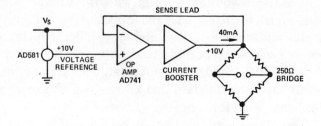

a. Basic bridge-drive circuitry

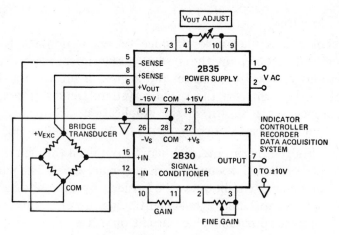

b. Use of transducer power supply for bridge drive

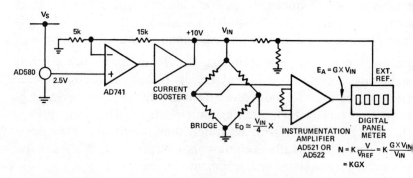

c. Basic ratiometric system using a low-cost 2.5V reference

Figure 2-5. Bridge excitation schemes

take many forms. While the early and elegant forms of analog microammeters and mirror galvanometers could easily resolve sub-microvolt variations, they are hardly applicable to today's industrial environment, which calls for considerably greater ruggedness and speed of response, and the ability to interface with analog or digital signal-handling circuitry. The most general and least troublesome, from the standpoint of circuit design and assembly, is to use packaged "system-solution" signal conditioning, which covers the gamut from card options associated with inteligent measurement-and-control systems* to packaged modules that include an adjustable-gain instrument amplifier, noise filtering, and excitation.†

However, even if one opts for packaged solutions, it is still worthwhile to know what the alternatives are. Furthermore, for reasons of performance, cost, overall system requirements, (or sheer satisfaction), a user may be inclined towards specific circuit options applicable to one's own system.

A simple and appealing circuit employing a single operational amplifier is shown in Figure 2-6. Though it maintains a voltage null across the bridge (like the circuit of Figure 2-4), current is not nulled; it has voltage gain, but the cost is high. The external resistances must be carefully chosen and matched to maximize common-mode rejection (CMR); the ideal case of everything equal, as shown in the figure, is hard to realize in practice. Also, it is difficult to switch the gain (and permit adjustments to maximize CMR), without a great deal of cost and trouble. Finally, depending on gain, the nonlinearity can be up to twice that of the bridge alone.

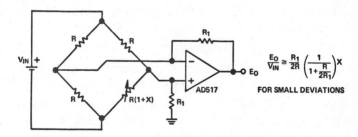

$$\frac{E_O}{V_{IN}} \cong \frac{R_1}{2R}\left(\frac{1}{1+\frac{R}{2R_1}}\right)X$$

FOR SMALL DEVIATIONS

Figure 2-6. Single op-amp as a bridge amplifier

Perhaps the most widely used form of amplifier for reading bridge outputs is the *instrumentation amplifier*. The instrumentation amplifier (unlike the general-purpose op amp) is a committed gain block, generally characterized by low drift, high common-

*For example, MACSYM 2
†For example, 2B31, with self-contained excitation

mode rejection, high input impedance, and the capability of maintaining specified performance over a range of gains, typically from 1 to 1000 (Figure 2-7). Gain is a function of the ratio of two resistances, which do not have circuit connections in common with the inputs; the gain can be adjusted by adjusting or switching the resistance ratio. Some devices require that both resistors be connected externally; some require only a single external resistance, and some contain all the necessary resistors for a number of standard gains and require only external programming by jumpers, switches, or digital logic.

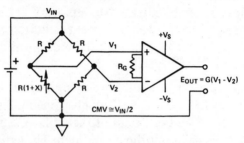

Figure 2-7. Differential-input instrumentation amplifier applied to bridge measurement

Common-mode errors and common-mode rejection will be discussed below and in the next chapter. When X is zero, the output should be zero. If the amplifier has been adjusted for zero output when V_{IN} = 0, then any error related to V_{IN} that appears at the output is known as a *common-mode-voltage error*. The ability of the amplifier circuit to minimize that error is known as *common-mode rejection* (CMR), a quantity expressed logarithmically in decibels. CMR is usually specified at 60Hz with 1kΩ source imbalance.

The instrumentation amplifier has a balanced differential input. This means that the output voltage is proportional to the difference between the input voltages; and the input terminals, which present a high impedance to the input source, are electrically similar. High common-mode rejection means that the amplifier is sensitive only to the *difference* between the input voltages, even if they are swinging over a wide range, and the difference is quite small. For example, if the common-mode input is a voltage swinging over a ±10V range, the difference is a 10mV signal, the gain is 1000, and 1mV of common-mode error is desired, the *common-mode rejection ratio* (CMRR), the ratio of signal gain to common-mode gain, must be 1000/0.0001 = 10 million. Expressed logarithmically, *common-mode rejection* (CMR) is 20 $\log_{10}$ (10)[7] = 140dB, in this case.

Because the instrumentation amplifier has differential inputs, it is useful as a readout of signals from transducers in wide variety, whether balanced (as bridges generally are) or unbalanced (single-ended, with a ground return having questionable stability). It is available in a variety of forms, ranging from monolithic amplifiers, to hybrids, to discrete modules; it is also at the heart of signal-conditioning modules.* With specifications such as $0.5\mu V/°C$ for offset temperature coefficient, 100–140dB common-mode rejection, and better than 0.01% nonlinearity, they can serve well in the majority of bridge-measurement applications. In very high-precision applications, problems arise when even these low levels of drift and common-mode error are excessive. Instrumentation amplifiers may also prove marginal or inappropriate in applications calling for high CMRR with potentially destructive high voltage under either fault or operating conditions.

It is possible to reconfigure circuits so as to improve one or more aspects of performance by simple level shifts. For example, Figure 2-8 shows how the use of split supplies can effectively reduce the dc common-mode voltage to zero. This is a neat trick which works well in many situations: but it does nothing to reduce drifts with time and temperature, and it provides no improvement with respect to dynamic common-mode variations.

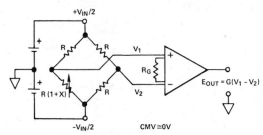

Figure 2-8. Using center-tapped supply to minimize common-mode voltage

An *isolation amplifier* is useful for applications where the bridge may be at a high potential with respect to the signal-conditioning circuitry, or where there must be no galvanic connections between the bridge and grounded instrumentation circuitry (for example in patient monitoring).

*An example of each: AD521, AD522, Model 610, Model 2B30 signal-conditioner

An isolation amplifier is one in which there is no galvanic path, and very low capacitance, between some combination of input, output, and power supply, hence no possibility for dc current flow, and minimal ac coupling. The usual combination isolates the input from both the power supply and the output (2-port isolation); but some devices have isolation between all three sections (3-port isolation). Isolation can, in concept, be provided via any means of energy transfer, from acoustics and lasers to magnetics and microwave. The easiest-to-implement and most widely used isolators today are electromagnetically coupled.

Figure 2-9 shows a basic configuration in which an isolation amplifier is providing the readout for a bridge. DC power is converted to high-frequency ac and coupled across the isolation

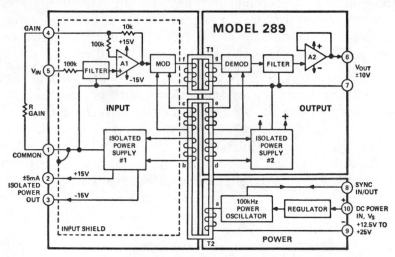

a. Block diagram of isolator

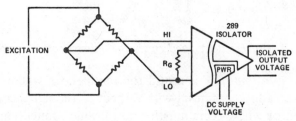

b. Typical application

Figure 2-9. Isolated bridge measurement. Gain is controlled by R_G. Bridge excitation may be low-frequency ac (transformer coupled) or dc.

barrier to the input section, where the ac is rectified to provide power for the input stage and for external isolated circuitry. The input signal modulates the ac carrier and is coupled to the output stage, where it is synchronously demodulated (using a sync signal coupled from the oscillator) and provided at the output.

The isolation results in extremely high common-mode rejection, with drift that is comparable to that of some instrumentation amplifiers. Because the amplifier's front end is truly floating, it is possible to withstand thousands of volts of common-mode voltage (CMV).

For the lowest drifts with time and temperature, amplifiers that employ *choppers* are used. Maximum drift, for modules such as the 261K, is $0.1\mu V/^\circ C$. A noninverting chopper amplifier is an op amp in which the dc offset between the feedback point and the input modulates a high-frequency carrier, which is then ac-amplified and demodulated, to produce an offset-corrected dc output level. Because of their internal circuit architecture, most chopper-type amplifiers are unbalanced single-ended-input devices; that is, the input and output voltage share a common terminal. This might appear to inhibit their ability to take a differential measurement across a bridge. However, if the bridge excitation is derived from a floating source, such as a battery or an isolated dc-to-dc converter, or an ac signal coupled via a transformer, then the output, though grounded, can be viewed as floating with respect to the excitation.

As Figure 2-10 shows, only the signal voltage appears at the amplifier input; as long as the supply floats, there is no common-mode voltage presented to the amplifier. Thus, this approach yields very low drift and high common-mode rejection. The remaining

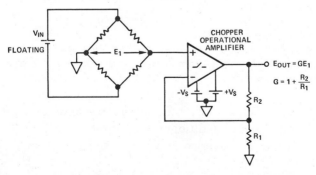

Figure 2-10. Chopper-stabilized bridge amplifier

source of concern is the accuracy and stability of the floating excitation source. A floating supply with regulation and stability *better than the required measurement accuracy* must be employed. Since the purpose of the chopper was to get high precision, it is reasonable to expect that the required measurement accuracy in such applications is high. The design and construction of highly stable floating power supplies, while certainly feasible, may be an unnecessarily high price to pay.

A better arrangement is shown in Figure 2-11, which provides very high accuracy, limited principally by the isolator's gain stability, at reasonable cost. Here a chopper amplifier is used as a freely floating preamplifier via an isolation amplifier. Power for the chopper is obtained from the isolation amplifier's isolated front-end supply (which is made available at a set of terminals on many models*). Since the preamplifier is isolated, it performs an essentially floating measurement of the bridge output, yet the bridge may be referenced to the same ground as the rest of the system. Because of this, a single grounded supply may be used to drive the bridge and also to provide a ratiometric measurement at the output (Figure 2-5c).

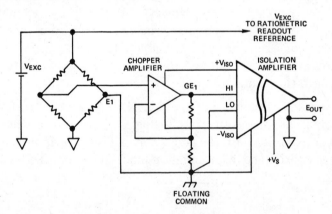

Figure 2-11. Chopper-stabilized, isolated bridge amplifier, using floating supply voltage from isolation amplifier front-end to power chopper amplifier

For most bridge applications, an instrumentation amplifier or a packaged signal conditioner will provide sufficient accuracy and resolution. However, it is worthwhile to be aware of the tools

*For example, Model 289

that are available for isolation, increased stability, and increased accuracy. Examples of applications in which they may be required are provided in the Applications section, in somewhat greater detail.

Bridges are perhaps the most pervasive element found in electrical measurement. This chapter has sought to provide a preliminary and practical acquaintance with their architecture, excitation, and readout. More information, in the more-general context of transducer interfacing, will be found in the chapters that follow.

Interfacing Considerations Interference

Chapter 3

In the first two chapters, it is easy for the reader to get the (correct) impression that transducer interface circuitry often involves the ability to deal with small signals. Whether one is simply seeking to resolve signals that are inherently small or to accurately handle signals with a wide dynamic range (or both!), one must pay attention to some factors that are not shown on a professionally drawn schematic diagram. These matters are sufficiently important to warrant discussion here before any further consideration of circuit elements, circuits, or system applications.

One can classify interference problems roughly in these three areas:

> *Problems generated locally* (e.g., unwanted thermocouples)
> *Problems communicated within a subsystem* (e.g., via grounds)
> *Problems originating in the outside world* (e.g., power-frequency interference)

Causes and cures of a few of the most-frequently encountered problems will be discussed here. Several useful publications (two of which are available from Analog Devices without charge) will be found in the Bibliography. They should be part of the working library of anyone concerned with low level signals or performing precision measurements in the face of great odds.

LOCAL PROBLEMS

In low-level measurements, strict attention must be given to the materials used (or found) in the signal path. As Figure 1-5b shows, the combinations of solder, wire, binding posts, etc., found in a simple circuit can generate substantial thermal emf's. Since they usually occur in pairs, it is useful to go to pains to keep such pairs at the same temperature, and in general, to seek to minimize thermal gradients by thermal shielding, heat-sinking, alignment along isotherms, and separation of high- and low-power circuitry. Even such innocuous practices as joining stranded wires from two different manufacturers can produce 200nV/°C, or twice the hard-won drift level of a chopper-stabilized amplifier.

Sockets permit convenient replacement of electrical components and subassemblies, but a poor one can introduce contact resistance, thermal potentials, or both, and can be a source of failure if exercised more than a few times. Because a good mechanical switch can be quite expensive, every effort should be expended to eliminate the requirement for high-performance switches by judicious choice of the point in the circuit at which switching takes place. For example, in an op amp circuit, gain-switching terminals should be wired to the high-level output, rather than to the summing point.

In general, the price for the convenience provided by switches, relays, connectors, and sockets is increased uncertainty of resolving low-level signals, i.e., somewhat poorer resolution and accuracy, increased noise, and lower reliability than might be attained in a direct-wired system. Since the electronics is a small element of cost in many of the ultimate applications for which transducers are required, it usually pays to obtain the best-quality hardware available. There are available specialized connectors designed for signal-carrying applications, with excellent electrical characteristics. Very often, the transducer manufacturer will specify or provide such a connector. Such switches as the Leeds & Northrop Type 34 are quite good with respect to contact potentials and resistance.

Banana jacks are the best connector generally available for low-level dc and are in common use in standards laboratories. For heavy-duty use, barrier strips with screw-type terminals are widely employed. It's not a good idea to use BNC-type connectors for low-level dc measurements, though they are optimal for high-frequency ac work.

As Figure 1-5b shows, even solder can become a culprit at low levels, because of the thermal emf's generated in solder joints. Special low-thermal solders, such as Kester Type 1544, should be used at the inputs to microvolt-level circuits. There are instances where it may be necessary to deliberately cut and re-splice (with solder) a connection in a line, in order to counterbalance a thermal emf introduced by a splice or solder connection in another line. Needless to say, one should seek to keep both connections at the same temperature.

SUBSYSTEM PROBLEMS

The lines drawn on a schematic drawing do not usually resemble the circuit wiring that is actually desired; Unlike the circuit wiring, the lines on a drawing are assumed to be free of resistance, inductance, and capacitance. Since they are, they may go to any lengths for purposes of clarity, pleasing appearance, or minimum drafting effort. For clarity, wires that go to many locations—such as power-supply leads, and "grounds"—may simply be terminated in tags, such as V_S, or $\bigtriangledown$; it is assumed that when the device is built, all such points will be at equal potential. But how equal is "equal" when we're processing signals in which microvolts may be significant? The two functions that suffer the greatest abuse due to lack of care in thinking and drawing are *grounding* and *bypassing*.

There is perhaps no more-misused word in electrical engineering today than "ground". Marconi and the Edison Company intended that grounded parts of the circuit be connected to rods that were literally "earthed", or driven into the ground. And today there are parts of electrical circuits for which the term "ground" is still validly used because of a solid tie to Earth (however, the electrical properties of Earth differ a great deal from place to place). The term has come to include a variety of other forms of common connection, including "grounding" to a metal chassis or housing, "grounding" to the low side of a power supply, and "grounding" to the common connection for input and output signals—in all cases, with very tenuous (if any) relationships with the Earth.

A consequence of the careless use of the "ground" concept can be seen in Figure 3-1. In (a), we see the lower end of several parts of a circuit, and the power supply, as drawn by the engineer, who intends that all the ground symbols be at the same potential. In (b), the circuit is redrawn, following the general configuration of (a), but with a (presumably equipotential) line connecting the "grounded" points, In (c), the circuit has been wired in approxi-

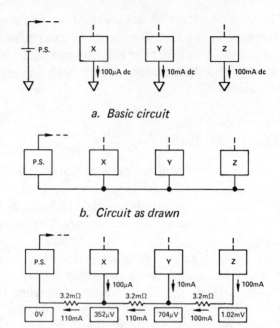

a. Basic circuit

b. Circuit as drawn

c. Circuit wired with 6" lengths of #18 wire. Voltages at each "ground" point are shown.

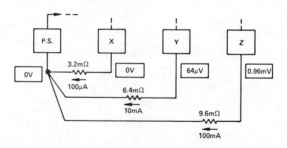

d. Circuit wired to single-point ground

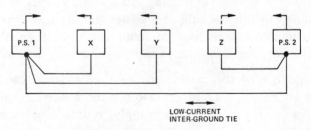

LOW-CURRENT
INTER-GROUND TIE

e. Separate supply for Z

Figure 3-1. A grounding example

mately the same configuration as (b) with #18 wire; the resistance of the 6″ lengths is about 3.2 milliohms (mΩ). The low end of circuit Z is at about 1mV, and the low end of low-power (and perhaps small-signal) circuit X is at 352μV. If, for example, X were an op amp, with its + input tied to the local ground point, the summing point would effectively be at an offset of 352μV with respect to a signal source referenced to the power supply. It is worth noting that the #18 wire specified here is perhaps somewhat heavier than the usual #20 or #22 wire used in electronic wiring; it would have even larger voltage drops.

One approach to improvement of the situation is shown in (d); a separate lead is run from each circuit to the low end of the power supply. The offset at circuit X is now negligible, the offset at circuit Y has been reduced by more than 90%, but the offset at Z is still about 1mV. Further improvement for Z, if necessary, could be gained by using heavier wire for its return, or perhaps by interchanging X and Z physically, if other considerations permit, or even by the use of a separate supply (e).

The circuit of (d) has essentially achieved the objective of (a), i.e., to return all of the circuit low points to a single common "ground" point and avoid the sharing of voltage drops in long leads. Note that each line is returned separately, and that no mixing of ground currents is permitted. In practice, the single-point ground may be an actual block of metal, to provide the lowest possible resistance at the common point.*

The common may be a *heavy* bus, just so long as interference is kept at a satisfactorily low level. Such a bus might be a suitable common connection for digital circuitry; it would then be connected to the analog common point to establish the basic system common.

Systems involving multiple power supplies and multiple chassis require more thought. Often, returning all lines, regardless of which supply they originate from, to the common point, and then tying the system common terminals together there will work. In other systems, such as (e), all +5V loads are returned to +5V common, all +15V loads returned to +15V common, and a final line is run between the common terminals to tie them together. In

---

*If power-supply voltage drops must be minimized, the "high" leads may be wired in a similar manner.

multiple-supply systems, intelligent experimentation may be needed to achieve the best compromise.

Digital ground lines are usually quite noisy and have large current spikes. All analog common leads should be run separately from the digital common leads and tied together at only one point (Figure 3-2).

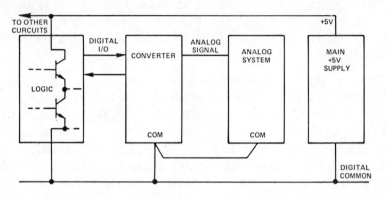

Figure 3-2. This connection minimizes common impedance between analog and digital (including converter digital currents)

In some cases, grounding problems can be solved by using instrumentation amplifiers as buffers at critical locations, converting the ground voltage into a common-mode voltage, which is rejected by the amplifier's differential input Figure 3-3. If very high-level and very low-level circuits must be used together, it may be desirable to float the circuits and transfer digital information via opto isolators and analog information via isolation amplifiers.

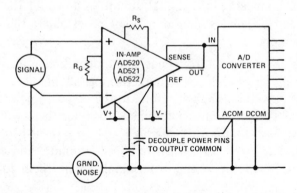

Figure 3-3. Use of instrumentation amplifier to interface separate ground systems

Besides the dc and low-frequency ground problems, there is also potential coupling of fast ac and transient signals from high-level circuits to low-level circuits through common power-supply and wiring impedances. Another way such signals can be coupled is a result of the fact that many internally compensated IC operational amplifiers do not have a dynamic connection to common; instead the reference for the output integrator is connected to one side of the power supply, and the output is subject to its perturbations, even if ground is as steady as a rock (Figure 3-4).

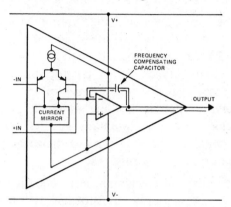

Figure 3-4. Typical op-amp circuit architecture.
Reference for output integrator is V–.

Both situations call for *bypassing* of high-frequency signals around slower analog circuitry by the use of judiciously placed capacitors. The capacitors are connected *directly* from amplifier power-supply terminals to the low-impedance common point.

Figure 3-5 illustrates this technique, as applied to decoupling the

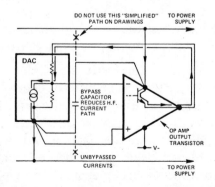

Figure 3-5. Bypassing power supplies for virtual-ground applications.
Arrows show unbypassed current flow.

noisy digital drive of a d/a converter from the analog output op amp. Note that if the bypass capacitor is drawn randomly as a straight connection between the two supply buses, and if it is physically connected in the same way, it will serve no useful purpose. Indeed, it can be harmful, by providing a low-impedance ac path for noise to go from a dirty ground to a previously clean V_S.

OUTSIDE AND LOCAL INTERFERENCE

AC signals at high power and high frequency can be coupled to low-level analog circuitry via stray capacitance and inductance. High dc voltage can be coupled to high-impedance input terminals via leakage conductance. Proper layout, shielding, and guarding are the defenses against these sources of interference. Elements of proper layout include keeping as much distance as possible between high-energy and low-energy circuits, and between digital and analog circuits, and as *short* a distance as possible between destinations of low-level wiring.

Both electrostatic and electromagnetic shielding are warranted in many cases. Power-supply transformer fields (especially from small modular supplies) are notorious sources of seemingly inexplicable problems. Simple approaches are either to use shielded types, or to mount the supply remotely from sensitive circuits. Paradoxically, the same circuit that requires the center tap of a transformer to be close by to achieve a high-quality grounding scheme may be disturbed by the intensity of the transformer's magnetic field. Often, it is necessary to experiment in order to determine the most suitable layout. For example, one can connect a simple R.F. choke across an oscilloscope probe to determine the presence and relative strength of high-frequency fields; this procedure will be found helpful in preliminary layout assessment.

Electrostatic and electromagnetic interference from the ubiquitous 50-60Hz power line can be brought under control by minimizing areas of wire loops, use of twisted pairs, shielding, and bandwidth limiting in low-frequency circuits. Battery-powered circuits, despite their isolation from the mains, are still susceptible to 60Hz pickup. A few picofarads of capacitance to ground can seriously degrade the performance of an ill-prepared battery-powered amplifier.

Often, small, shielded houses will have to be constructed around an amplifier or an assembly to prevent interference from fields or energy-radiation within the system itself. Another form of shielding that must be considered is *thermal* shielding against hot spots and gradients. Consider the relationships between placement of high-dissipation and low-level portions of a system. Make sure components are not exposed to drafts and unequal temperature distributions.

Guarding is a technique used to prevent ac and dc leakage currents from degrading circuit performance. Guards are conductive surfaces usually placed close to inputs or other points in a circuit that are sensitive to stray currents. The guard is driven at low impedance to a potential that is very close to (nominally equal to) the voltage being protected. If the guard is properly placed stray leakage current will be absorbed by the guard. Since the guard is at the same potential as the protected point, no leakage current will flow between them. Figure 3-6a shows a simple application of a guard to a high-impedance unity-gain follower input. Another advantage of the guard is in reducing the input capacitance of the circuit. Stray capacitance from the outside world is to the guard, not the

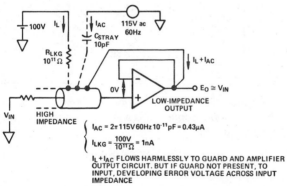

$$I_{AC} = 2\pi\,115V\,60Hz\,10^{-11}pF = 0.43\mu A$$

$$I_{LKG} = \frac{100V}{10^{11}\Omega} = 1nA$$

$I_L + I_{AC}$ FLOWS HARMLESSLY TO GUARD AND AMPLIFIER OUTPUT CIRCUIT. BUT IF GUARD NOT PRESENT, TO INPUT, DEVELOPING ERROR VOLTAGE ACROSS INPUT IMPEDANCE

Figure 3-6a. Example of guard circuit—follower-connected op amp

input terminal; again, since the guard and the input terminal are at about the same potential, very little displacement current flows between them, hence, the effective capacitance is virtually nil. In circuit-board work, a guard ring should completely enclose the protected point, and a jumper should be used for interconnections (Figure 3-6b). This technique is particularly important at dc when low bias-current op amps are used.

Additional discussions of the necessity and techniques for reducing appropriate forms of interference will be found in descriptions of a number of applications in the Applications section.

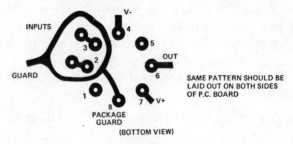

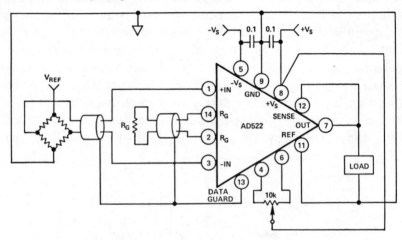

b. Board layout for guarding inputs of AD515 with guarded TO-99 package

c. Use of the AD522 instrumentation amplifier's guard terminal to guard both the input connections and connections to a remote gain-setting resistor

Figure 3-6. Guarding

ANALOG FILTERING

Despite the best of efforts to reduce interference, the signal may include too much noise to ignore. Indeed, this may not be the fault of the designer of the signal-conditioning circuitry; the noise may have been present in the signal itself, or if the input information is modulating an ac signal, demodulation and filtering must take place. For most of the transducers discussed in this book, the information is fairly slow, with typical bandwidths of the order of 10Hz maximum; filtering of these signals is relatively easy.

Again, it is important to observe that no filter, however well intended or crafted (or cheap) is justifiable as a substitute for proper attention to wiring, layout, and shielding techniques; rather, it is an adjunct to them. Every effort should be expended to keep noise out of the system. Low-level leads (especially those at high impedance levels) should be shielded or guarded and run away from noise-generating sources. Such sources can include motors, transformers, fluorescent and carbon-arc lamps, induction heaters, and—in general—any source of electromagnetic radiation. It is poor practice to allow any form of noise into a system *carte blanche* on the premise that "filtering will take care of it." Once the interface system has been tightened against noise, the filtering may be added to clean up any remaining undesired signals, especially those present in the original signal.

Besides reducing noise, filtering is also used to reduce bandwidth so that any frequency components at greater than 1/3 to 1/2 the sampling rate in sampled-data systems are negligible, to prevent aliasing.

In this chapter, we shall describe some easily understandable and accessible filtering techniques. For those interested in deeper understanding of filter design, references to useful sources of information are provided in the Bibliography. For those interested in treating filters as "black boxes", names of manufacturers and their specialties may be found in industry Buying Guides. Filters are quite often found in commercially available signal conditioners.*

For the applications discussed here, the most useful form of filtering is the *low-pass* filter, which responds perfectly to very low-frequency signals and has a great deal of attenuation at high frequencies. Since the phenomena we are concerned with are primarily low-frequency phenomena and much of the noise that is picked up, whether line-frequency, radio frequency, or carrier frequency, tends to be at higher frequencies, this form of response is widely used. The simplest low-pass filter consists of a resistor and a capacitor. Figure 3-7 shows three ways (active filter circuits) in which it is commonly implemented. The first of these, (a), can be used wherever the output of the filter drives a high-impedance

---

*For example, the Analog Devices 2B30 and 2B31 signal-conditioning modules contain 3-pole Bessel filters with 2Hz cutoff frequency. Higher cutoff frequencies are available by the use of 3 external resistors. The *modified Bessel* filter's amplitude and phase response are designed for optimum time-domain step response: fast rise time and minimum overshoot.

load, such as the input of an instrumentation amplifier or an op-amp connected as a follower. (b) and (c) are inverting-op-amp versions; c shows how two independent time constants can be achieved, forming a second-order low-pass filter.

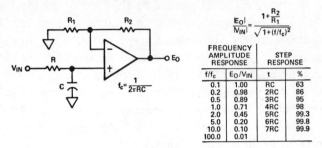

$$\frac{|E_O|}{V_{IN}} = \frac{1+\frac{R_2}{R_1}}{\sqrt{1+(f/f_c)^2}}$$

$$f_c = \frac{1}{2\pi RC}$$

FREQUENCY AMPLITUDE RESPONSE		STEP RESPONSE	
f/f_c	E_O/V_{IN}	t	%
0.1	1.00	RC	63
0.2	0.98	2RC	86
0.5	0.89	3RC	95
1.0	0.71	4RC	98
2.0	0.45	5RC	99.3
5.0	0.20	6RC	99.8
10.0	0.10	7RC	99.9
100.0	0.01		

a. First-order RC low-pass filter unloaded by follower. Gain is set independently of time constant (RC).

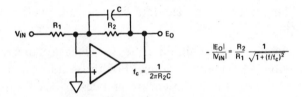

$$-\frac{|E_O|}{V_{IN}} = \frac{R_2}{R_1}\frac{1}{\sqrt{1+(f/f_c)^2}}$$

$$f_c = \frac{1}{2\pi R_2 C}$$

b. Inverting op amp as first-order RC filter. Gain is set by R_2/R_1, time constant by R_2C. For independent setting, fix R_2, adjust C for cutoff frequency, R_1 for gain.

$$-\frac{|E_O|}{V_{IN}} = \frac{\frac{R_3}{R_2+R_1}}{\sqrt{1+(f/f_{c1})^2}\sqrt{1+(f/f_{c2})^2}}$$

$$f_{c1} = \frac{\frac{1}{R_1}+\frac{1}{R_2}}{2\pi C_1}$$

$$f_{c2} = \frac{1}{2\pi R_3 C_2}$$

c. Adapting (b) as a second-order filter with two independently adjustable time constants

Figure 3-7. Simple RC active low-pass filters

Figure 3-8 shows the basic circuit architecture for second- and third-order low-pass filters, using a single follower-connected op amp. By tailoring the values of capacitance and resistance, a variety of response characteristics can be achieved, for a given cutoff frequency, e.g., three equal time constants, maximal flatness in the pass band, lowest phase shift, fastest settling time to a given

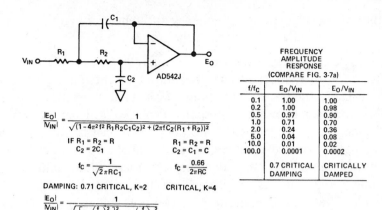

$$\frac{|E_O|}{|V_{IN}|} = \frac{1}{\sqrt{(1 - 4\pi^2 f^2 R_1 R_2 C_1 C_2)^2 + (2\pi f C_2 (R_1 + R_2))^2}}$$

IF $R_1 = R_2 = R$ $R_1 = R_2 = R$
$C_2 = 2C_1$ $C_2 = C_1 = C$

$$f_c = \frac{1}{\sqrt{2}\pi R C_1} \qquad f_c = \frac{0.66}{2\pi R C}$$

DAMPING: 0.71 CRITICAL, K=2   CRITICAL, K=4

$$\frac{|E_O|}{|V_{IN}|} = \frac{1}{\sqrt{\left[1 - \left(\frac{f}{f_c}\right)^2\right]^2 + K\left(\frac{f}{f_c}\right)^2}}$$

f/f_c	E_O/V_IN	E_O/V_IN
	FREQUENCY AMPLITUDE RESPONSE (COMPARE FIG. 3-7a)	
0.1	1.00	1.00
0.2	1.00	0.98
0.5	0.97	0.90
1.0	0.71	0.70
2.0	0.24	0.36
5.0	0.04	0.08
10.0	0.01	0.02
100.0	0.0001	0.0002
	0.7 CRITICAL DAMPING	CRITICALLY DAMPED

a. 2nd order filter—0.7 critically damped and critically damped

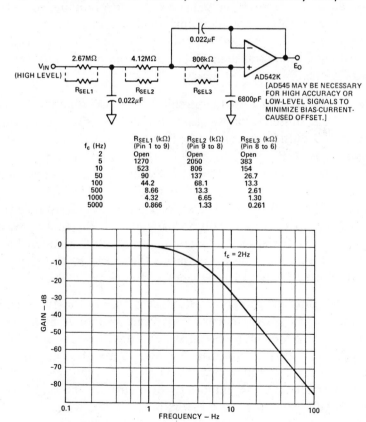

f_c (Hz)	R_SEL1 (kΩ) (Pin 1 to 9)	R_SEL2 (kΩ) (Pin 9 to 8)	R_SEL3 (kΩ) (Pin 8 to 6)
2	Open	Open	Open
5	1270	2050	383
10	523	806	154
50	90	137	26.7
100	44.2	68.1	13.3
500	8.66	13.3	2.61
1000	4.32	6.65	1.30
5000	0.866	1.33	0.261

b. 3rd order filter—modified Bessel response

Figure 3-8. 2nd and 3rd order RC low-pass active filters

input step level, minimum frequency difference for a given accuracy in the pass band and rejection in the stop band.

Low-pass filter responses may have many *poles* (corresponding to the *order*, or number of energy storage elements having a lagging effect), and the transfer functions of cascaded filters may be multiplied together to produce a response typical of a higher-order filter. Conversely, a complicated transfer function can be factored into unit lags and second-order responses. The choices of time constants, natural frequencies, and damping for these filter elements determine the composite response of the filter. For example, a 3-pole filter (the equivalent of a two-pole and a one-pole) may be combined with a two-pole filter to form a 5-pole filter, which will have an ultimate rolloff rate of 100dB per decade. Typical standardized sets of filter parameters (corresponding to distribution of poles in the complex domain—but enough of that!), such as Butterworth, Bessel, Chebyshev, Paynter, etc., provide fast rolloff, aperiodic transient response, linear phase shift with frequency, etc.

The values indicated for the circuit in Figure 3-8b are for a modified 3-Pole Bessel response (used in the 2B31 signal conditioner) with a 2Hz cutoff frequency (the frequency at which the amplitude response is 3dB down, to 71% of the low-frequency value). Its amplitude response can be seen to be rolling off at 60dB per decade (i.e., down by a factor of 1000 for every tenfold increase in frequency). The table indicates a set of 1%-tolerance resistance values that are connected externally in parallel with the filter's resistors to modify it for several commonly used cutoff frequencies. To predict the nominal response, move the curve to the right until the response is –3dB at the new cutoff frequency.

Low-pass filtering should be used as close to the front end as possible in a system. Theoretically, one can filter at any stage of a linear system—but not all systems are linear. If any noise is present ahead of a nonlinearity, its effects may be hard to get rid of at a later stage. Filter cost is minimized if the capacitance can be kept small; but the amplifier's bias current and any leakages develop voltage across the resistors in the filter, which establishes the upper limit to the resistance. Capacitors should generally be high-quality units having low leakage and low dielectric hysteresis—polystyrene, Teflon, mica, and polycarbonate are typical useful materials for the smaller sizes. For high capacitance, tantalum is a good choice. If electrolytic types are used, one should be careful about polarity.

If, despite the best efforts of the designer, interference of a single frequency (e.g., line frequency) is present, a narrow-band-elimination (*notch*) filter may be desirable. A popular configuration is the twin-tee, shown in Figure 3-9. The values given are ideal component values for rejecting 60Hz interference. The notch frequency and depth are sensitive to component values; the components should be well-matched and trimmed. The highest frequency of interest in the signal should be considerably lower than the notch frequency, since the notch is broad. This filter is generally used with a low-pass filter to minimize transmission of noise at higher frequencies, since the notch filter has unity gain at high frequency. The notch can be made sharper by feedback[1] but will require components having considerably greater stability.

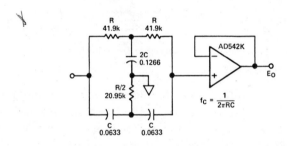

Figure 3-9. Twin-tee notch filter tuned to 60Hz. Typical components might be TRW MARG resistors and Arco type PT capacitors.

Another type of filter that is a useful building block is the *state-variable* filter. It consists of integrators in a feedback loop solving a differential equation. For example, a second-order state-variable filter has two integrators. Enough amplifiers are included to provide damping, sign inversion, and linear combinations to make use of this filter's versatility, since it can be used as a high-pass, low-pass, bandpass, or band-elimination filter, with adjustable Q, depending on how it is connected. It can also be used as an element of a complex multi-pole active-filter cascade. And it can be used with coefficients controlled by analog multipliers or multiplying DAC's

[1] Jung, Walter G., *IC Op-Amp Cookbook*, Howard W. Sams & Co., Inc., 1974, pages 338-340

to provide externally programmable parameters. An example of such a filter is shown in Figure 3-10.

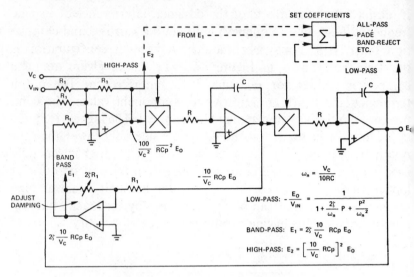

Figure 3-10.   Second-order state-variable filter with multiplier-controlled natural frequency

Because the foregoing filter blocks employ op amps for unloading or driving or for actual contributions to the filter transfer function through feedback, they are known as *active* filters. *Passive* filtering (resistors, capacitors, inductors in circuits requiring no external energy source) is used for fairly simple transfer functions, and where either amplification will be provided or the impedance and signal levels are sufficient to enable the output of the filter to drive the subsequent circuit elements (such as analog meters, recorders, etc.)

Specialized Filter Circuits

A simple way of significantly reducing both noise and high-frequency common-mode errors in instrumentation-amplifier circuitry is the use of a capacitor connected between the differential instrumentation amplifier's input terminals. This is useful in two ways. First, with the source resistances of the inputs, it serves to form an RC filter for the differential signal. Second, it tends to reduce any unbalance between the common-mode capacitances from both sides of the input to ground because of its cross-coupling effect. Figure 3-11 is a example that illustrates how it works.

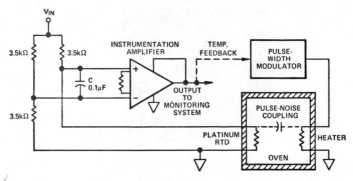

Figure 3-11. Reduction of unbalanced pulse noise in a temperature-monitoring system by the use of a shunt capacitor at the instrumentaion-amplifier input.

Here, a differential instrumentation amplifier is used to monitor the temperature of an oven. The platinum sensor resides within the oven, while the other resistors that make up the bridge circuit are located near the amplifier on the circuit board. The oven uses a pulse-width-modulator-driven transistor switch to control the power into the heater. While this circuit is efficient, it also generates noise, which is picked up by the platinum sensor, due to its proximity to the heater. The problem is aggravated by the inductance of the sensor, which is wirewound. Pulse noise could lead to erroneous readings in a monitoring DPM or multiplexed data-acquisition system.

Figure 3-12a shows the waveforms that are seen at the two inputs of the instrumentation amplifier. The transducer side of the bridge picks up substantially more noise than does the resistance divider (note that the resistance-divider trace has a 10X more sensitive scale factor).

In the absence of the capacitor, the amplifier's output reflects this unbalance (b). The lower trace shows the amplified difference between the two input signals (upper trace), which includes the amplified noise waveform.

With the $0.1\mu F$ capacitor connected between the amplifier inputs, the two input traces, corresponding to (a), appear very nearly equal, due to the cross-coupling effect of the capacitor for the noise. That they are indeed equal can be seen at the output of the amplifier (d); the noise excursion is well below 1mV (note the scale for the lower trace, compared to the scale for the lower trace in (b).

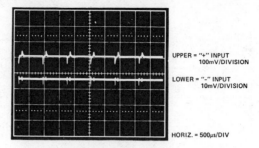

a. Input traces, no capacitor

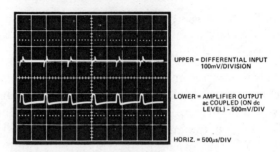

b. Differential input, and output— no capacitor

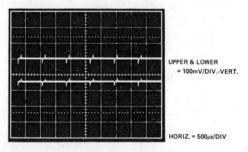

c. Input traces with capacitor

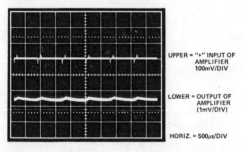

d. Common-mode input and output trace with capacitor. Note scale for lower trace.

Figure 3-12. Waveforms in circuit of Figure 3-11

On occasion, unusual filtering problems may arise. For example, consider the case of an instrumentation amplifier and a filter that are shared by the channels of a multiplexed system, which provides programmable gain. If each channel has noise of a known different character, the filter time constant may also be programmable in order to minimize scan time without duplicating hardware.[2] A highly simplified block diagram of such an arrangement is shown in Figure 3-13. In this way, signals having large amounts of low-frequency noise can have sufficient filtering and adequate settling time without slowing down the acquisition of relatively clean signals with mostly high-frequency noise.

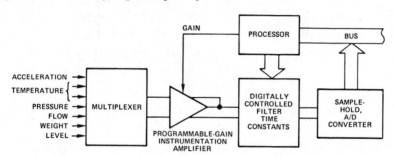

Figure 3-13. Data acquisition system using digitally controlled software-programmed filter

Another unusual application calls for a *derivative-controlled low-pass filter*. This is a filter that settles rapidly in response to a step change of input, then assumes a long time-constant for filtering out the low-level noise. Figure 3-14 depicts the use of such a filter in a rapidly-responding infant weighing scale where the constant motion of the baby would preclude stable high-resolution readings. Details can be found in the Applications section, Figures 15-8 through 15-10.

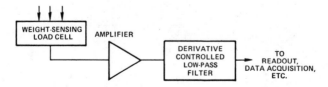

Figure 3-14. Derivative-controlled low-pass filter in a weighing application

[2] Several digitally programmable low-pass filters are described in the Analog Devices *Application Guide for CMOS Multiplying D/A Converters*, 1978.

Synchronous filters, utilizing sample-hold techniques, may be used to filter out noise in an interface. In some systems, noise is a direct consequence of the operating characteristics of components within the system. Switching power supplies, inductors, current surges, etc., often contribute noise which is both characterizable and predictable with respect to frequency and amplitude. In low-level circuitry, transients caused by such noise sources can cause saturation and slow recovery.

One way to minimize the effects of such noise sources is to use a sample-hold circuit in the signal path. The device is commanded to *hold* just before the spike is expected; when the noise is expected to have decayed to an acceptable level, the circuit is allowed to return to the *track* (normal operating) mode. This kind of circuit, which operates in synchronism with the noise bursts, is very similar to "deglitching" circuits used in fast digital-to-analog converters.

In the circuit of Figure 3-15, a synchronous filter is used to gate out the noise caused by ignition of an oil burner. The same command that instructs the oil-burner ignitor is used to gate the sample-hold circuit. During the time of the ignition pulse, the output of the interface remains at the level it had just prior to ignition. After ignition, the interface data is again back to real time. This technique is especially useful where the predictable noise occurs at relatively low duty cycles with respect to the desired system time frame, because the sample-hold action does

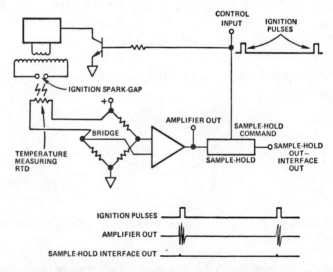

Figure 3-15. Synchronous filtering application

create "holes" in the data stream (but generally smaller and more-uniform ones than would occur if the spikes were allowed to communicate with the low-level circuitry).

Selecting a Cutoff Frequency
It is highly desirable to select as low as possible a cutoff frequency for the transducer interface filtering. The farther away the noise frequencies are from the frequency of interest, the easier it is to filter them out. A good guideline is to select a cutoff frequency which is as low as the desired information from the transducer will permit. For example, if nothing of interest occurs at frequencies above 1Hz, the 3-pole Bessel filter of Figure 3-8b, with a cutoff frequency (–3dB) of 2Hz, will permit measurements of reasonable fidelity at 1Hz, yet have attenuation of 70dB (3000X) at 60Hz.

On the other hand, piezoelectric accelerometers can generate information at considerably higher frequencies, and a higher noise cutoff frequency will usually be appropriate. In addition, circuits employing piezoelectrics (e.g., charge amplifiers) also require attention to *high-pass filtering* to minimize errors due to amplifier bias currents and the high noise gain at low frequencies due to the integrating effect of charge amplifiers.

In multi-transducer systems, where time-domain multiplexing techniques are used to scan between the various transducers, a critical look should be taken at the scan frequency's relationship to transducer frequency response. The scan frequency should be fast enough so that it captures changes occurring at the fastest rate of interest for a given transducer. And the dwell time should be long (short) enough to permit adequate time for settling without discernible changes of data. While the Nyquist criterion must be observed (sampling at about 3—or more—times the highest frequency of interest, and filtering-out of all higher frequencies to avoid aliasing), it may be foolishly redundant to scan a slowly-varying signal thousands of time per second. A useful technique is to group slow and fast signals so that all are scanned at rates appropriate to the actual rate at which their data can change.

Amplifiers and Signal Translation

Chapter 4

Transducers usually require amplifiers or related devices for buffering, isolation, gain, level translation, and current-to-voltage or voltage-to-current conversion. Most of these functions can be (and are) performed by operational amplifiers. However, the level and character of the circuit techniques required for best design and implementation may well cause the prudent system designer to seek packaged "system solutions."

In this chapter, we will briefly discuss the basics of op-amp circuits and device selection.* Instrumentation and isolation amplifiers, which were mentioned briefly in Chapter 2, will be discussed further here, and there will be a brief treatment of signal conditioners, level translators, and other subsystems. Practical aspects of these devices are generally discussed in relation to actual applications in the Applications section.

OPERATIONAL AMPLIFIERS

The operational amplifier (op amp†) is a high-gain amplifier designed for use in feedback circuits to perform stable predictable *operations*, which are inherently determined by the external components and configuration rather than by the amplifier's open-loop gain magnitude.

Figure 4-1 shows the symbol of a *differential operational amplifier* and its basic open-loop response. Since the gain, A, is ideally infinite

---

*There are many good sources of information on the theory and application of op amps. Several that we have found valuable are listed in the Bibliography.
†An *operational amplifier* is any circuit configuration, wherever found, that behaves in the manner described here. The term *op amp* is an abbreviation for a *device* manufactured for use in operational-amplifier applications.

and usually quite high (10^5 to 10^6 at low frequencies), the differential input signal must be nearly zero for the output to be within the limits.* If an operational amplifier is *ideal* (infinite gain, bandwidth, and input impedance, zero input offset voltage and bias current, zero output impedance), a number of useful consequences may be inferred. Usually, a real amplifier can be found with sufficiently ideal parameters for use in a given application based on postulating an ideal amplifier.

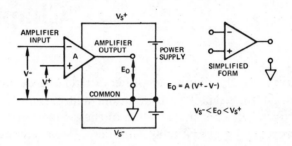

a. Differential operational amplifier circuit, open-loop

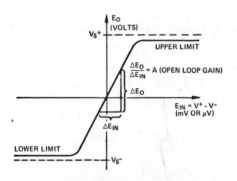

b. Plot of output vs. input for circuit in (a). Note that input scale is millivolts or microvolts.

Figure 4-1. Op amp basic circuit and response

Since the output must be whatever is necessary to make the input voltages, V^+ and V^-, nearly equal via feedback, the op amp is an excellent unity-gain *follower* (Figure 4-2a). If the feedback is attenuated, we have a *follower with gain* (b), where the gain is

---

*Most op amps may also be used as *comparators,* where the output is always at one limit or the other, depending on the polarity of the input signal (beyond the threshold of linear operation).

accurately set by a resistance ratio. If a metering resistor, R_M, is connected from the negative input to common, then the output must provide a current through the non-grounded load element, Z_L, precisely equal to V_{IN}/R_M (and independent of the nature of Z_L), in order to maintain the inputs at equality (c).

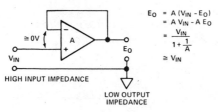

a. Unity-gain follower

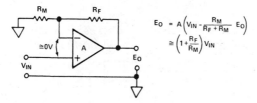

b. Follower with gain has high input impedance, low output impedance, gain depends only on R_F/R_M

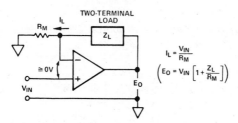

c. Current generator—I_L depends only on V_{IN} and R_M; it is independent of Z_L and E_O

Figure 4-2. The op amp as a follower

If the plus input is connected to common, the output voltage must be whatever is required to hold the negative input at zero (*virtual ground*). Connecting a resistance in series with an input voltage will cause a current to flow (through the non-grounded element Z_L) equal to V_{IN}/R_M (Figure 4-3a). If Z_L is a resistance, R_F, the output voltage, E_O, must be $-(R_F/R_M) V_{IN}$, a polarity-inverted voltage that may be greater than, less than, or equal to V_{IN} in magnitude, depending only on the resistance ratio (b). If a second voltage, V_S, is applied to the negative input via a re-

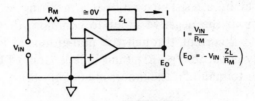

a. Current source with input at virtual ground. Current depends only on V_{IN} and R_M, is independent of Z_L or E_O

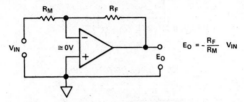

b. Op amp as inverting amplifier. Gain depends only on R_F/R_M and may be greater than, less than, or equal to unity

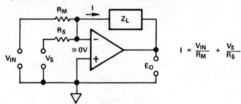

c. Op amp as summing current source. Currents are independent of one another and of E_O or Z_L

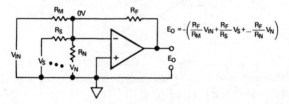

d. Op amp as summing amplifier. Resistance ratios determine individual gains or attenuations independently

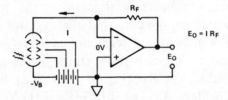

e. Op amp as current-to-voltage converter. Photomultiplier tube sees zero impedance—an ideal load. Output voltage is scaled by R_F.

Figure 4-3. Inverting applications of an op amp

sistor, R_S, the current through Z_L will depend on the sum of the two independent input currents, V_{IN}/R_M and V_S/R_S; the currents are independent because the input resistors appear to the input voltages to be grounded (c). Since the currents are summed in R_F, the output voltage depends on the independently weighted sum, $V_{IN}(R_F/R_M) + V_S(R_F/R_S + \ldots V_N(R_F/R_N)$ (d). An input current source (such as a photomultiplier tube) can be connected directly to the negative input (an ideal zero-impedance load); its current is summed and transconducted to a voltage (e).

Since the current through R_M depends only on V_{IN} and the nature of R_M, R_M could be replaced by a capacitor (to form a differentiator, $I = CdV/dt$), by an element having a nonlinear voltage-current relationship (e.g., a diode in series with a resistor), or by an n-terminal network of linear and nonlinear impedances, with a number of inputs (Figure 4-4a); the resulting current through R_M would depend only on the short-circuit property of the input circuit. That current could be measured by a voltage via R_F and the amplifier output, without disturbing the value of the current. Similarly, the feedback network could be dynamic, nonlinear, and/or an n-terminal network; for an input current deter-

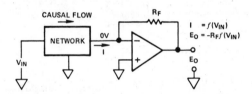

a. Output voltage is an arbitrary function of V_{IN}, determined by the nature of the input network, whether purely resistive, fixed or dynamic, linear or nonlinear, constant or time-varying

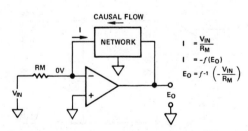

b. The output voltage is an inverse function of V_{IN}. The amplifier causes the output current of the network to be equal to $-V_{IN}/R_M$ by the manipulation of E_O

Figure 4-4. Direct and inverse functions using an op amp

mined by V_{IN} and R_M, the output would have to be whatever voltage is necessary to maintain that current (b). For example, if R_F were replaced by a capacitor, the circuit would be an integrator; if R_F were shunted by a capacitor, the circuit would have a unit-lag response: if the forward drop of a diode were being tested, the diode could be the feedback element—the output voltage would be equal to the diode drop at a given current determined by V_{IN}/R_M.

In the differential mode, an op amp will function as an analog adder, subtractor, and multi-input adder-subtractor (Figure 4-5a). A subtractor with gain can be used as an instrumentation amplifier, if care is taken to null out common-mode errors due to the amplifier, resistive elements, and input loading. The subtractor-with-gain can be combined with two cross-coupled followers-with-gain to form the circuit used in many conventional instrumentation amplifiers (b). The common-mode rejection of the subtractor is increased by the gain of the input followers, and the signal source is unloaded by the high input impedance of the followers.

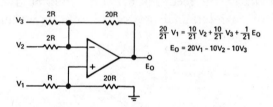

$$\frac{20}{21} V_1 = \frac{10}{21} V_2 + \frac{10}{21} V_3 + \frac{1}{21} E_O$$

$$E_O = 20V_1 - 10V_2 - 10V_3$$

a. Adder-subtractor circuit. Simple rules for computing resistor ratios for any number of positive and negative inputs and any combination of gains or attenuations can be found in Analog Dialogue 10-1 (1976), page 14

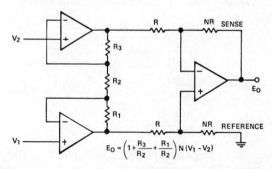

$$E_O = \left(1 + \frac{R_3}{R_2} + \frac{R_1}{R_2}\right) N (V_1 - V_2)$$

b. Classical 3 op-amp instrumentation amplifier configuration

Figure 4-5. Differential op-amp configurations

There are many useful applications of op amps, both singly and in groups; indeed, the op amp has become the most versatile of elements in the circuit-designer's kit. This brief review has sought to establish the basic elements that constitute this versatility (and if pondered can suggest a great many more applications):

> High gain and low input excursion
>
> Negative input terminal must follow positive input terminal
>
> Output voltage must be whatever value (within limits) is necessary to achieve this.
>
> Op amp can generate inverse functions (e.g., gain is equal to inverse of attenuation of feedback circuit).
>
> Op amp can convert current to voltage, voltage to current.
>
> Op amp can establish virtual ground: voltage null and current balance.
>
> Op amp can sum inputs independently.
>
> Op amp can function as sign inverter.
>
> Op amp activates passive circuitry, linear or nonlinear, static or dynamic, single-element or network.
>
> Differential signal handling

Differences between op amps

Although the "ideal op amp" doesn't exist, op amps have been designed to optimize one or another set of parameters that are crucial in a given family of applications. The titles of selection charts in an op-amp catalog listing[2] provide some indication of the groupings of parameters that are commonly required:

1. General purpose (lowest cost)
2. FET-input, low bias current
3. Electrometers (lowest bias current)
4. High accuracy, low-drift differential
5. Chopper amplifiers (lowest drift)
6. Fast wideband
7. High output
8. Isolated op amps

For transducer interfacing: categories 1, 2, and 4 are the most relevant; 3 & 5 are required frequently; 8 is needed often (and will be discussed in the Isolator section); 6 is not commonly re-

---

[2] *Data-Acquisition Products Catalog*, Analog Devices, Inc.

quired; and 7 includes a potpourri of special requirements, often arrived at most economically by a proprietary design or an external device. The key specifications are *input offset*, *offset drift* with time, temperature, and power-supply voltage, *input bias current* and its drift, and *common-mode rejection*. Open-loop gain, bandwidth, noise, rated output voltage and current, stability with feedback and capacitive loading, etc., are all relevant (often critical) considerations, but the key specifications are usually the starting point in narrowing the universe in which the search for the "best choice" is conducted.

Offset is a small voltage, from tens of microvolts to millivolts, that appears—in effect—in series with the inputs of an operational amplifier (Figure 4-6); it acts just like a small voltage applied to the + input and is amplified in the same way. The principal cause of offset is mismatches in the input differential gain stage. A set of terminals is usually provided for connecting a potentiometer for adjusting the offset to zero, in an analogy to the zero-adjustment of a mechanical scale. If the amplifier's offset is substantial compared to the voltage representing the phenomenon being measured, these terminals should be used for nulling the offset. What is "substantial" depends on the application. Sometimes, a 5% error can be tolerated; in other cases, 0.01% is excessive. The amplifier's offset-adjustment terminals should be used *only* for adjusting the amplifier's voltage offset to zero. They should not be used to compensate for other errors in the amplifier or for offsetting the input signal itself; adjustment of offset will unbalance the amplifier, usually causing increased temperature sensitivity; if additional adjustments are made to correct other errors, the errors with temperature may become intolerable.

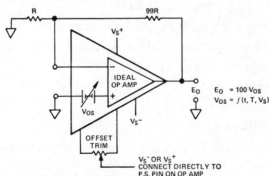

Figure 4-6. Offset voltage of an operational amplifier, and circuit for measuring it directly

The offset voltage is not absolutely stable. It drifts with time, temperature, and power-supply variations. This last is not usually a source of concern, since well-regulated power supplies are easy to buy. Drift with temperature (*tempco – temperature coefficient*) ranges from *tenths* of microvolts per °C to *tens* of microvolts per °C.

The drift error is specified as an average over a range (or ranges) of amplifier operating temperatures. Drift is usually nonlinear with temperature, and the worst drifts are generally at the extremes. For this reason, somewhat better drift performance may occur at mild temperatures ("room temperature"), but it is risky practice to place absolute reliance on this effect to save small amounts of money.

Amplifiers at constant temperature tend to drift with time in random-walk fashion. The drift *is* random and rarely—if ever—accumulates (at least, for a well-constructed device). The magnitude of drift expected over a year is $\sqrt{12}$ times the drift per month.

For the greatest stability with time and temperature, *choppers* are used. A chopper is simply a switch, exercised at a rapid rate, that compares the amplifier input with the desired null. Any offset produces a square wave, which may be ac-coupled without drift, amplified, then demodulated synchronously to form a low-frequency (dc) voltage, which is an amplified version of the offset. This dc is then fed back to provide an offset reduction proportional to the net gain. Chopper amplifiers have narrow bandwidth and tend to be single-ended; they are used in two ways, either as an adjunct to an inverting op amp for obtaining reduced drift, while maintaining the full bandwidth of the main amplifier (*chopper-stabilized op amp*), or as a narrow-band non-inverting high-input-impedance amplifier (*chopper op amp*).*

Bias current is the dc current that must flow at the input terminals of an op amp; in some cases it is a leakage current, in other cases it is the actual base current required by bipolar transistors for transistor bias. It flows through the external circuitry of the op amp and develops voltage across resistors or rates-of-change of voltage in capacitors ("open circuits"). For example, a bias current of 10nA (10^{-8}A) flowing from the negative input terminal of an op amp through a 1MΩ feedback resistor produces a 10mV output

---

*Analog Devices Models 235 and 261 are respective examples of the two categories.

offset. If the amplifier is a unity-gain inverter, this offset is equivalent to a 5mV voltage offset. In many types of differential-input op amps (*but not all*), the bias currents at the + and − inputs tend to be equal and to track one another; this permits first-order error reduction if the effective resistances at the two inputs are made equal, usually by insertion of a resistor of appropriate value in series with a low-impedance input (Figure 4-7).

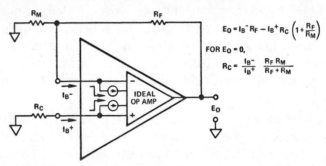

$$E_O = I_B^- R_F - I_B^+ R_C \left(1 + \frac{R_F}{R_M}\right)$$

FOR $E_O = 0$,

$$R_C = \frac{I_B^-}{I_B^+} \cdot \frac{R_F R_M}{R_F + R_M}$$

Figure 4-7. Bias-current contributions to amplifier output offset (V_{OS} nulled). If currents are matched and tracking, voltage can be nulled if effective parallel resistances at the two inputs are equal

For low bias current, field-effect transistors (FET's) are used at the amplifier inputs; bias current for FET-input op amps ranges from 75fA (75×10^{-15}A) max for the AD515L to 50pA max for the AD542J, while for bipolar op amps it ranges from 1nA max (AD517L) to 500nA max (AD741).The lowest bias currents are found in parametric op amps (using varactor choppers*), 10fA max in Models 310/11†. Bias current changes by about 1%/°C (decreasing with increasing temperature) in bipolar op amps, doubles every +10°C in FET's and doubles every +7°C in parametrics.

Common-mode rejection affects an op amp's ability to deal with common-mode signals, when the positive input is active. Examples are found in follower circuits (measuring potentiometer outputs) and differential-amplifier circuits, (measuring bridges). An ideal operational amplifier responds only to the *difference* voltage between the inputs (normally very small, as noted in Figure 4-1), independently of whether the inputs are at +10V, −10V, or zero

---

*See *Analog Dialogue*, Volume 1, No. 3 for an explanation of how amplifiers can achieve low bias current using the parametric approach.

†Modules—they should not be confused with quite different IC's having nearly the same designations, but with the prefix, "AD": viz., AD310 and AD311.

(with respect to the power-supply common). However, due to slightly different sensitivities of the + and − inputs, or variations in offset voltage as a function of common-mode level, some common-mode voltage may appear at the output. If the output error voltage, due to a known magnitude of common-mode voltage (CMV), is referred to the input (dividing by the gain-for-a-voltage-at-the + input, or *closed-loop gain*), it reflects the equivalent *common-mode error* (CME) voltage effectively in series with the inputs.

Common-mode rejection ratio (CMRR) is defined as the ratio of common-mode voltage to the resulting common-mode error voltage (referred to the input). Common-mode rejection (CMR) is often expressed logarithmically:

$$\text{CMR (in decibels)} = 20 \log_{10} \text{CMRR}$$

For CMRR = 20,000, CMR = 86dB.

The precise specification of CMR is complicated by the fact that the common-mode voltage error can be a nonlinear function of common-mode voltage (and also varies with temperature). As a consequence, CMR data published by Analog Devices are average figures, assuming an end-point measurement over the common-mode range specified. The incremental CMR about small values of CMV may be greater than the average CMR specified (and smaller in the neighborhood of a large CMV). Published CMR specifications for op amps pertain to very low-frequency voltages, unless specified otherwise; CMR decreases with increasing frequency (often characterized graphically on op-amp data sheets).

When op amps are used singly or in groups to form instrumentation amplifiers to measure input difference voltage at arbitrary common-mode levels, the amplifier's common-mode error is only one source of CME; in addition, one must consider the effects of impedance-ratio mismatches in the feedback circuit and loading of the source by the amplifier's input circuit. For such applications, it is a good idea to consider using a committed *gain block*, which incorporates all required circuit elements (except perhaps for the gain-determining element, which must be at the user's option and may be fixed, variable, or programmable). Such gain blocks, available in IC and modular form, and also as input elements of data-acquisition systems, are known as *instrumentation amplifiers*. When an instrumentation amplifier's gain is programmed by digital logic, it is known as a *Programmable-Gain-Amplifier,* (PGA).

INSTRUMENTATION AMPLIFIERS

An instrumentation amplifier is a committed "gain block" that measures the difference between the voltages existing at its two input terminals, amplifies it by a precisely set gain—usually from 1V/V to 1000V/V or more—and causes the result to appear between a pair of terminals in the output circuit. Referring to Figure 4-8,

$$V_S - V_R = G (V^+ - V^-)$$

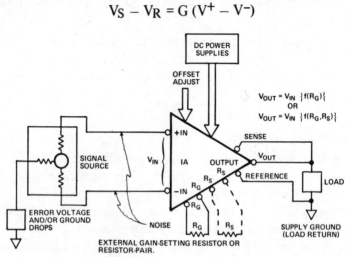

Figure 4-8. Basic instrumentation amplifier functional diagram

An ideal instrumentation amplifier responds only to the *difference* between the input voltages. If the input voltages are equal ($V^+ = V^- = V_{CM}$, the *common-mode voltage*), the output of the ideal instrumentation amplifier will be zero. The gain, G, is described by an equation that is specfic to each Model.

An amplifier circuit which is optimized for performance as an instrumentation-amplifier gain block has high input impedance, low offset and drift, low nonlinearity, stable gain, and low effective output impedance. Example of applications which capitalize on these advantages include the interfacing of thermocouples, strain gage bridges, current shunts, and biological probes; preamplification of small differential signals superimposed on large common-mode voltages; signal-conditioning and (moderate) isolation for data acquisition; and signal translation for differential and single-ended signals wherever the common "ground" is noisy or of questionable integrity.

Instrumentation-amplifier modules, IC's, and front-ends are usually used in preference to user-assembled op-amp circuitry because they offer optimized, specified performance in low-cost, easy-to-use compact packages. For applications that call for very high common-mode voltages or isolation impedances, with galvanic isolation, the designer will choose Isolators (to be discussed).

As Figure 4-8 shows, instrumentation amplifiers have two high-impedance input terminals, a set of terminals for gain-setting resistance(s)—except for those units that have digitally programmable gain—and a pair of feedback terminals, labeled "sense" and "reference", as well as a set of power-supply and offset-trim terminals.* When the *sense* (V_S) feedback terminal is connected to the output terminal and the *reference* terminal (V_R) is connected to power common, the output voltage appears between the output terminal and power common.

The V_S and V_R terminals may be used for remote sensing—to establish precise outputs in the presence of line drops; they may be used with an inside-the-loop booster follower to obtain power amplification without loss of accuracy; and they may be used to establish an output current that is precisely proportional to the input difference signal. A voltage applied to the V_R terminal will bias the output by a predetermined amount. It is important always to maintain very low impedance (in relation to the specified V_S and V_R terminal input impedances), when driving the V_S and V_R inputs, in order not to introduce common-mode, gain, and/or offset errors. In some devices, the V_R terminal is used for fine adjustments to common-mode rejection and/or offset.

Specifications

Although all specifications are relevant and none should be neglected, the most-important specs in transducer interfacing are those relating to *gain* (range, equation, linearity), *offset, bias current*, and *common-mode rejection*.

Gain range is the range of gains for which performance is specified. Although specified at 1 to 1000, for example, a device may work at higher (and, in the case of the AD521, *lower*) gain, but performance is not specified outside that range. In practice, noise and drift may make higher gains impractical for a given device.

*Some units include a *guard* terminal, which follows the common-mode voltage. It may be used to drive a guard conductor to reduce unbalanced noise, leakage, and capacitance (see *guarding*, Chapter 3).

Gain equation error or "gain accuracy" specifications describe the deviation from the gain equation when R_G is at its nominal value. The user can trim the gain or compensate for gain error elsewhere in the overall system. Systems using digital processing can be made self-calibrating, to take into account the lumped gain errors of *all* the stages in the analog portion of the system, from the transducer to the a/d converter. *Gain vs. temperature* specifications give the deviations from the gain equation as a function of temperature.

Nonlinearity is defined as the deviation from a straight line on the plot of output vs. input. The magnitude of linearity error is the maximum deviation from a "best straight line", with the output swinging through its full-scale range, expressed as a percentage of full-scale output range.

While initial *voltage offset* can be adjusted to zero, shifts in offset voltage with time and temperature introduce errors. Systems that involve "intelligent" processors can correct for offset errors in the whole measurement chain, but such applications are still relatively infrequent; in most applications, the instrumentation amplifier's contribution to system offset error must be considered.

Voltage offset and *drift* are functions of gain. The offset, measured at the output, is equal to a constant plus a term proportional to gain. For amplifier with specified performance over the gain range from 1 to 1000, the constant offset is essentially the offset at unity gain, and the proportionality term (or slope) is equal to the change in output offset between $G = 1$ and $G = 1000$, divided by 999. To refer offset to the input, divide the total output offset by the gain. Since offset at a gain of 1000 is dominated by the proportional term, the slope is often called the "RTI offset, $G = 1000$". At any value of gain, the offset is equal to the unity-gain offset plus the product of the gain and the "RTI offset, $G = 1000$".

The same considerations apply to offset drift. For example, if the maximum RTI drift is specified at $25\mu V/°C$ at $G = 1$, $2\mu V/°C$ at $G = 1000$, it will be $(23/G + 2)\mu V/°C$ at any arbitrary gain in the range. At the output, the corresponding drift will be $(23 + 2G)$ $\mu V/°C$. Voltage offset as a function of power-supply voltage is also specified RTI at one or more values of gain.

Input bias currents have the same causes as in op amps. They can be considered as sources of voltage offset (when multiplied by the source resistance). For balanced sources, the *offset current*, or difference between the bias currents, determines the bias-current contribution to error. Differences between the bias currents with temperature, common-mode level, and power supply voltage can lead to voltage offset or common-mode error.

Although instrumentation amplifiers have differential inputs, there must be a return path for the bias currents, however small. If the path is not provided, those currents will charge stray capacitances, causing the output to drift uncontrollably or to saturate. There-fore, when amplifying the outputs of "floating" sources, such as transformers and thermocouples, as well as ac-coupled sources, there must be a dc "leak" from both inputs to common*. If a dc return path is impracticable, an *isolator* must be used.

Common-mode rejection, in instrumentation amplifiers, is a measure of the change in output when both inputs are changed by equal amounts. CMR is usually specified for a full-range com-mon-mode voltage change, at a given frequency, and a specified imbalance of source impedance (e.g., 1kΩ source imbalance, at 60Hz). The common-mode rejection ratio (CMRR) in instrumen-tation amplifiers is defined as the ratio of the signal gain, G, to the ratio of common-mode signal appearing at the output to the input CMV. In logarithmic form, CMR (in dB) = 20 log$_{10}$ (CMRR). Typical values of CMR in instrumentation amplifiers range from 70dB to 110dB. In the high-gain bridge amplifiers found in modu-lar signal-conditioners†, the minimum line-frequency common-mode rejection is of the order of 140dB.

ISOLATION AMPLIFIERS

The isolation amplifier, or *isolator*, has an input circuit that is galvanically isolated from the power supply and the output circuit. Isolators are intended for: applications requiring safe, accurate measurement of dc and low-frequency voltage or current in the presence of high common-mode voltage (to thousands of volts) with high common-mode rejection; line-receiving of signals transmitted at high impedance in noisy environments; and for safe-ty in general-purpose measurements where dc and line-frequency leakage must be maintained at levels well below certain mandated

---

*This consideration, often neglected, is perhaps the most frequent cause of phone calls for help from our application engineers.
†Analog Devices 2B30 and 2B31

minima.* Principal applications are in electrical environments of the kind associated with medical equipment, conventional and nuclear power plants, automatic test equipment, industrial process-control systems, and field-portable instrumentation.

While, in concept, any non-conducting medium may be used for

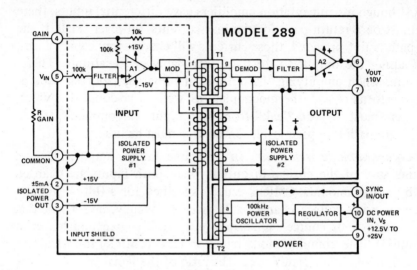

a. Block diagram

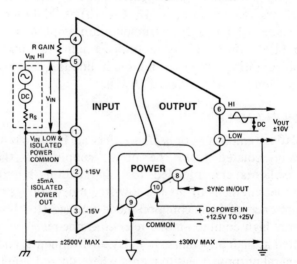

b. Simplified functional diagram

Figure 4-9. Typical isolation-amplifier configuration

*Examples of such requirements may be found in UL STD 544 and SWC (Surge Withstand Capability) in IEEE Standard for Transient Voltage Protection 472–1974.

isolation, including light, ultrasonics and radio waves, the medium that is currently in widest use, because of its low cost and (relatively) easy implementation, is transformer-coupling of a high-frequency carrier for communicating power to and signals from the input circuit. Figure 4-9 is a diagram of a *3-port isolator*, one in which the power source, the front end, and the output circuit are all isolated from one another.

The two-wire primary dc power source provides excitation for a high-frequency oscillator. The oscillator output is coupled across the isolation barrier to the input section, providing power for the front end and for external isolated accessories (such as preamplifier circuitry). The input signal is amplified and modulates the carrier; the modulated waveform is coupled across the isolation barrier to the output section, where it is demodulated by a phase-sensitive demodulator (using the oscillator as a reference), and filtered.

The amplifier in this example is a resistor-protected op amp (actually, the protection works both ways—it protects the amplifier against differential overloads and it protects sensitive input sources from supply voltage if the amplifier malfunctions), connected for a programmable gain from 1 to 100 volts/volt, as determined by a single external resistor.

Since both input terminals are floating, the amplifier functions effectively as an instrumentation amplifier. Because of the transformer coupling, the output of these devices is isolated from the input stage.

The isolated power-supply output terminals can be used to provide floating power to transducers, preamplifiers, and other circuitry within the current limitations of the supply. Examples of the various ways of using isolation amplifiers in transducer interfacing will be found in the Applications section.*

The device pictured in Figure 4-9, typical of Model 289 and similar types, is a completely self-contained device. However, isolators are also available in several other useful forms. For example, if there are many input channels to be isolated, economies can be realized by the use of a common oscillator, which has the additional benefit of making it possible to avoid the possibility of

---

*Additional information on isolation and instrumentation amplifiers is to be found in the *Isolation and Instrumentation Amplifier Designers' Guide,* available from Analog Devices, Inc., at no charge.

small errors due to beat frequencies developed by small amounts of crosstalk. In any event, isolators intended for use in multichannel systems should be *synchronizable.*

Another type of isolated amplifier, of interest for many applications, is the *isolated operational amplifier*.* This device has an operational-amplifier front end, which may be used to perform the many operations of which op amps are capable (including integration, differentiation, summing, etc.) and provide an output that is isolated from the input.

Specifications of Isolation Amplifiers

The key specifications of isolation amplifiers are of two kinds: performance specifications and isolation specs. The key performance specifications are similar to those for instrumentation and operational amplifiers, with a few additions: Nonlinearity, CMR — inputs to outputs, CMR — input to guard, offset voltage referred to input, input noise. Key isolation (and protection) specs include: maximum safe differential input, CMV — inputs to outputs, leakage current, overload resistance.

CMR — inputs to outputs indicates ability to reject common-mode voltages between the input and output terminals. It is important when processing small signals riding on high common-mode voltages.

CMR — input to guard indicates the ability to reject differential voltage between the low side of the signal and the guard. It should be considered in applications where the guard cannot be directly connected to the signal low.

Maximum safe differential input is the maximum voltage that can be safely applied across the input terminals. It is important to consider it for fail-safe designs in the presence of high fault voltages.

CMV — inputs to outputs is the voltage that may be safely applied to both inputs with respect to the outputs or power common. It is a necessary consideration in applications with high CMV input or when high-voltage transients may occur at the input.

Leakage current is the maximum input leakage current when power-line voltage is impressed on the inputs. It is a vital consideration for patient safety in medical applications.

Overload resistance is the apparent input impedance under con-

---

*Model 277 is an example of this type.

ditions of amplifier saturation. It limits differential fault currents.

The family of isolated devices is a growing one. Besides today's isolation amplifiers, isolated op amps, and isolated power supplies (dc-to-dc converters), there are isolated *voltage-to-current converters**, *thermocouple signal-conditioners†*, *high- and low-level multiplexers§*, and *d/a converters‡*, with yet more to come.

SYSTEM SOLUTIONS

Beyond the basic elements, such an op amps, instrumentation amplifiers, and isolators there are families of increasingly committed devices, which take more and more of the interfacing burden from the shoulders of the system designer. Since this volume is concerned with making the burden lighter for those who desire to bear it, we will mention system solutions only briefly, with the suggestion that complete information on the topics covered can be had merely by inquiring in the appropriate quarters.

Complete signal conditioners have already been mentioned. Several applications of these devices, which incorporate excitation, bridge amplification, and filtering, will be found in the Applications section.

Microcomputer interface cards‖ are available in great number to interface with the most-popular microcomputers. *Input* cards respond to current inputs, and single-ended or differential high-level voltage sources, providing multiplexing, programmable-gain amplification, sample-hold, conversion, and enough memory-mapped software compatibility to make programming easy. *Output* cards provide the choice of voltage output or 4-20mA current output, for a number of channels (typically four). Compact *combined I/O* cards provide much of the capability of both types. Although low-level signal-handling and transducer interfacing must be dealt with off-board, these devices solve the problem of interfacing the medium-to-high-level analog signal to a microcomputer without tears.

---

*For example, Model 2B22 0 to +10V to 4-to-20mA V/I converter, 1500V dc continuous isolation

†For example, Model 2B50 isolated (±1500V), cold-junction-compensated, 12 bit-accurate, direct-connected (screw terminal) thermocouple signal conditioners

§For example, Models 2B54 and 2B55 Low-level (with gain) and high-level isolated 4-channel multiplexers (±1000V dc CMV, 12-bit accuracy, 400 channel/s scan speed), companion cold-junction compensator: 2B56

‡For example, DAC1423 10-bit isolated DAC with 4-20mA output: bus compatible, bumpless transfer, total power from loop supply, manual control backup, 1500V isolation

‖The RTI-1200 real-time-interface family

MACSYM 2 Measurement And Control SYsteM is designed to bring the entire job of interfacing under the control of the system user. Its function cards (in wide variety) permit direct wiring of signals to and from the transducers. From that point on, the amplification, signal conditioning, data acquisition, display, control, and programming are unified in an easy-to-understand-and-use subsystem, with its own keyboard, fast, flexible multitasking MACBASIC programming language, and interactive display. A block diagram and photograph of the system can be seen in Figure 4-10.*

A compatible system, MACSYM 20, has the same transducer interfacing capability as MACSYM 2, with somewhat less intelligence, at less than half the cost of MACSYM 2.

*MACSYM 2 is described in a comprehensive overview in *Analog Dialogue* 13-1. A description of MACSYM 20 can be found in *Analog Dialogue* 14-1.

a. Front view of MACSYM 2, showing keyboard and display

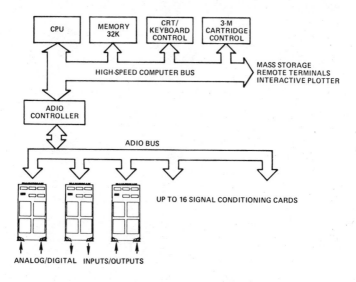

b. Block diagram of MACSYM 2

Figure 4-10. MACSYM 2 measurement and control system

Offsetting and Linearizing

Chapter 5

In the earlier chapters, we have discussed the basics of electrical-output transducers, signal-conditioning for bridges, interference problems, and the characteristics of amplifiers used in signal conditioning. This chapter deals with two processes that are often found useful in making sensitive and accurate measurements: *offsetting* and *linearizing*.

OFFSETTING

As used here, *offsetting* embraces the use of analog techniques to shift the level of a signal by a predictable amount. Typical applications include:

> Measurement of small changes about a large initial value
> Incremental measurements, employing a device that has an absolute scale (gauge vs. absolute pressure, °C vs. K)
> Reducing a common-mode level
> Restoring or introducing an offset (for example, in converting from a 0 to +10V range to a 4 to 20mA range for transmitting analog signals).

For some applications, an accurately developed constant is unnecessary. For example, an isolation amplifier can be used to measure small differential signals riding on large common-mode voltages. The isolator simply ignores the common-mode voltage. Or if the useful portion of the transducer output is an ac signal, capacitive or transformer coupling may be used to eliminate the dc level.

Figure 5-1 provides an illustration of offsetting with a *bridge* configuration. The transducer, in this case, is a linear voltage divider. The variable we are interested in is the fractional deviation from half-scale, $k \dfrac{E_b}{2}$.

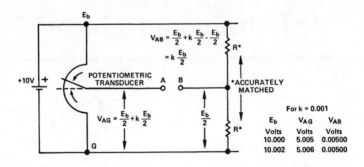

Figure 5-1. Use of bridge in offsetting. Note that, while the fractional part of V_{AG} changes by 20%, V_{AB} is essentially unchanged.

If E_b is 10V, then $E_b/2$ is 5V, and a 0.1% deviation, k, is repre sented by a 5-millivolt change. An accurate digital voltmeter a AG, between the wiper of the potentiometer and the low end o E_b, would read +5.005V, and the positive deviation, +0.005V could be read by ignoring the initial digit. Unfortunately, the output is highly sensitive to error; a 0.02% error in either the power-supply voltage or the meter reading would cause a *20%* *error* in the measured value of $k \dfrac{E_b}{2}$.

If, on the other hand, a bridge were formed, using an accurately matched resistance pair to produce an offsetting voltage, V_{BG}, and the meter were to read the *difference*, $V_{AB} = 0.005V$, the effec of small changes of power-supply voltage would be cancelled because both voltages would be changed by about the same amoun (to within *microvolts*, in the example). Meter accuracy could be improved by the use of a low-full-scale voltage range (e.g., 0.199 volts on a 3 1/2-digit meter) and/or preamplification of the difference voltage (which is much easier to handle than the sum o a large and a small voltage).

Users of resistive transducers that do not inherently involve bridge find this bridge-building technique quite useful. Examples of sucl transducers include thermistors, RTD's, and strain gages.

Another—less-critical—form of offsetting is used in reducin; measurements from absolute to gauge, from kelvin to degree

Celsius, or in translating the outputs of high-level transducers that have offset ranges. The technique is simple input summation in operational or instrumentation amplifiers, or using the *reference* input of an instrumentation amplifier. Figure 5-2a shows how the $1\mu A/K$ absolute-reading AD590 output may be scaled to $1mV/K$ by resistance and offset by a fixed $273.2mV$ to provide a voltage output representing °C. Figure 5-2b shows how the output of a high-level semiconductor pressure transducer, with an output range of 2.5 to 12.5V, representing a 0 to 100psi a range of pressure, might be offset to provide zero volts out for zero psi.

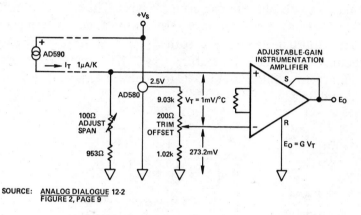

SOURCE: <u>ANALOG DIALOGUE</u> 12-2
FIGURE 2, PAGE 9

a. Offsetting the AD590 1µA/K temperature transducer for Celsius measurements

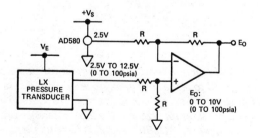

b. Offsetting semiconductor pressure-transducer output

Figure 5-2. Range offsetting

Cold-junction compensation is a special form of offsetting used with thermocouples in applications where it is inconvenient to provide an ice bath for the reference ("cold") junction (the vast majority of applications fall within this category). The off-setting circuit measures the ambient temperature at the cold junction and adds a voltage approximately equal to the voltage expected to be developed by the cold junction but of opposite polarity. The net output of the circuit is equal to the Seebeck voltage of the measuring junction.

The circuit of Figure 5-3 provides cold-junction compensation (CJC) for a type J thermocouple (iron-constantan). The reference junction is established in intimate thermal contact with an AD590 temperature-to-current integrated circuit (current in microamperes is equal to absolute temperature, kelvin). The resistance network adds a constant term and a term proportional to temperature; when adjusted to read the correct value of voltage output at a nominal reference temperature (say 25°C), the circuit provides accuracy to within about 0.5°C for ambient temperatures between 15° and 35°C.

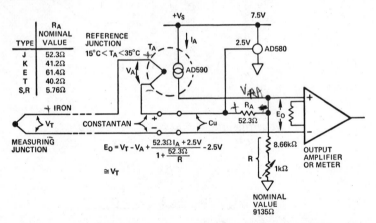

Figure 5-3. *Cold-junction compensation circuit for type J thermocouple, employing AD590 to sense cold-junction ambient temperature and provide compensation over the ambient range 15°C to 35°C. Voltage across R_A compensates for V_A, and the 2.5V reference offsets the voltage across R, due to AD590's 273.2µA (absolute temperature) measured at 0°C.*

The principal contributions to error come from the temperature coefficients of the voltage reference and resistances. While this example is for J thermocouples, the circuit will serve for other thermocouple types, if different resistance values are calculated and substituted for R_A.

Nominal values of R_A for some common thermocouple types are tabulated in Figure 5-3.

As noted in an earlier chapter, cold-junction compensation is not necessary if the ambient temperature variation is small compared to the temperatures being measured. Cold-junction compensation is provided in a variety of system-solution products.*

4-TO-20mA CURRENT TRANSMISSION

In many process-control applications, information is transmitted in the form of current, with a span of 16mA full scale, and an offset range of 4 to 20mA. The transmission of current provides a degree of noise immunity, since the received information is unaffected by voltage drops in the line, stray thermocouples, contact voltage or resistance, and induced voltage noise. At the same time, the offset provides a distinction between *zero* (represented by 4mA) and *no information*, due to an open circuit (zero current flow).

An additional benefit of this form of transmission is that, for some applications, power can be furnished remotely, via the 4mA of current that is not needed for information transfer. Thus, power is transmitted in one direction—signal in the return direction. No local source of power is needed at such a transducer, and only two wires are needed for transmission.

Finally, information in the form of current permits several loads at differing locations to be connected in series—up to a specified maximum voltage. For example, the output of a transducer could drive a chart recorder and a meter, and provide an input to a controller. Figure 5-4a shows a typical 4-to-20mA process-control loop using a modular voltage-to-loop-current converter. Figure 5-4b shows an isolated voltage-to-4-to-20mA converter; excitation for the output current is furnished (in this instance) by an external loop supply (but is also available within the unit).†

*Examples of products available from Analog Devices include the 2B50 and 2B51 thermocouple signal conditioners (2B50 has ±1500V dc isolation); the 2B56 cold-junction compensator, as a companion to the 2B54 low-level four-channel isolated multiplexer; the AD2036 6-channel scanning digital thermometer; and the thermocouple input cards provided with MACSYM intelligent Measurement And Control SYsteMs. An IC thermocouple preamplifier with cold-junction compensation will become available during 1980.

†The 2B20 meets the requirements for Type 3, Classes L and U (non-isolated), and the 2B22 meets the requirements for Type 4, Class U (isolated), of ISA Standard S50.1, "Compatibility of Analog Signals for Electronic Industrial Process Instruments."

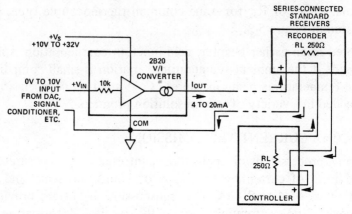

Figure 5-4a. Typical 4-to-20mA process-control loop using the 2B20 as a transmitter.

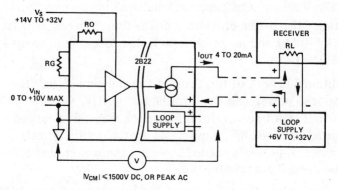

Figure 5-4b. Typical isolated 4-to-20mA process-control loop using the 2B22 as a modulator.

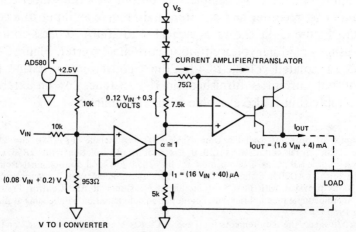

Figure 5-5. Basic 0 to 10V to 4 to 20mA translation circuit

The devices shown in Figure 5-4 represent low-cost system solutions, embodying specified guaranteed performance. However, there are a number of ways for a determined designer, familiar with op-amp circuitry, to build a successful voltage-to-current converter, using published op-amp circuits. Attention must be paid to the requirements for accuracy, grounding, input voltage range, output current and compliance voltage, power-supply requirements, reference, and dc stability. The rudiments of such a circuit are shown in Figure 5-5.

NONLINEARITY AND LINEARIZING – BASIC IDEAS

A linear system or element is one for which cause and effect are proportional; if there are several inputs, the output is proportional to their (weighted) sum*. Nonlinearity is a measure of the departure from proportionality. Figure 5-6 shows an input-output plot of an ideal linear relationship, a nonlinear relationship, and the difference between the two (*nonlinearity*).

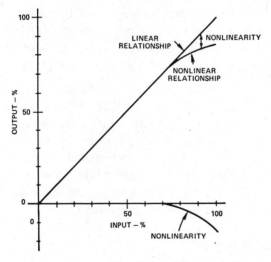

Figure 5-6. Typical nonlinear relationship

All devices are characterized by nonlinearity. For some, it is a desired characteristic (for example, the logarithmic characteristic of a log-antilog amplifier); for some others, it is a performance limit that is predictable by the nature of the device characteristics (thermocouple, off-null bridge). for the rest, it is a performance

*Graffito in an MIT facility: "Happiness is when it's linear."

limit that differs from device to device and must simply be accepted as a worst-case specification (usually in some such terms as the *maximum deviation from a "best straight line" over a given range of input*).

In this section, we are concerned with the second class, *undesired but predictable nonlinearities*. We shall discuss several means of dealing with them to obtain useful data.

Digital Linearizing

If the data are to be digitized and processed digitally, as soon as possible, it probably makes sense to perform any needed linearization in the digital domain. The principal techniques involve read-only-memory (ROM) and computational algorithms.

ROM has the fastest access and is used where processing capability and time are limited, and where the nonlinearity is well-defined and fixed. A ROM may be hard-wired to the converter output, and gated by the end-of-conversion flag, so that the signal presented for further processing has already been linearized. Each output level from the converter corresponds to an address in ROM, and the word stored at that address is either the correctly linearized value of the variable or an additive correction term.

If the nonlinear input source is to be looked at infrequently and memory is limited, but rapid mathematics is available, a mathematical function that approximates the inverse of the nonlinear relationship—or the difference between the ideal signal and the actual signal—can be derived and stored in program memory. Then, whenever the input from the nonlinear source is needed, the processor computes the correct value, based on its mathematical relationship to the measured input variable.

Analog Linearizing

For many applications, it is best to linearize the output of the transducer at some point in the analog process. This is obviously true for the case where no digital processing is used, but it is also true where limited processing capability and/or memory are available, and where the analog processing can be done simply and at low cost.

To linearize the signal from a transducer, one can—in some cases—modify the transducer circuitry; more often, some form of analog processing of the transducer output signal is used. An example of modifying the transducer circuitry can be seen in Figure 2-4, wheıe an amplifier provides a feedback signal that balances the bridge to obtain an output that is proportional to the resistance change of the active element. In another example, the active leg of a bridge may be driven by a current derived from the voltage applied to the reference leg. And yet another example of modifying the transducer circuitry is the use of networks involving thermistors and resistors to obtain an electrical output that is linear over limited ranges. Manufacturers use a technique involving resistors and combinations of differing thermistor elements to provide linearized devices having linearities to within 0.2°C over ranges such as 0° to 100°C.

If resistance is connected in series or in parallel with a simple thermistor, the output of the circuit can be linearized in a rudimentary way over limited ranges of temperature. Figures 5-7 and 5-8 illustrate the technique.

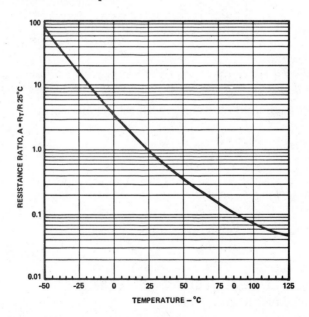

Figure 5-7. Typical thermistor characteristic. Note its essentially logarithmic character.

Figure 5-7 is the logarithmic plot of resistance ratio* vs. temperature for one type of thermistor. Although quite sensitive to tem-

*Like resistors, thermistors are available in a range of specified values of resistance (at 25°C), having similar normalized characteristics.

perature (resistance increases by 350% as temperature decreases from 25°C to 0°C), it is also highly nonlinear (approximately logarithmic). Yet it can be used in a simple circuit that has a sensitivity of about 1%/°C of the applied voltage over a range of about 50°C, with linearity to well within 5% of the range. Figure 5-8 shows the circuit and a plot of normalized output voltage vs. temperature.

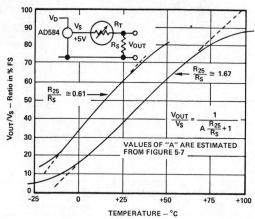

Figure 5-8. Calculated examples of linearized thermistor output, for device with the response of Figure 5-7, over limited ranges.

The approach used here starts with the relationship:

$$\frac{V_{OUT}}{R_S} = \frac{V_S - V_{OUT}}{A \cdot R_{25}},$$ (5.1)

whence

$$\frac{V_{OUT}}{V_S} = \frac{1}{A\,\dfrac{R_{25}}{R_S} + 1}$$ (5.2)

where

V_{OUT} is the output voltage of the resistive divider into a high-impedance load

V_S is the constant input voltage

A is the ratio of thermistor resistance at any temperature (R_T) to its resistance at 25°C (R_{25})

R_{25} is the specified resistance of the thermistor at 25°C, when operated at a sufficiently low power to avoid significant dissipation.

R_S is the value of series compensating resistance.

With the reasonable assumption that, over a limited range of temperature (T the absolute temperature, kelvin),

$$A \cong \alpha \, \epsilon^{\beta/T} \qquad (5.3)$$

and with values of A picked off the curve at two temperatures, ($0°$ and $50°C$), the values of α and β for that region were calculated to be about 1.44×10^{-6} and $4016K$.

If (5.3) is plugged into (5.2) and differentiated with respect to temperature*, and if different values of the ratio, R_{25}/R_S, are tried at various temperatures in the range, $0°C$ to $25°C$, calculations show that at a ratio of about 0.61, the derivative will exhibit little change, implying that the function is nearly linear.

Two curves are plotted in Figure 5-8. One is based on the value of $R_S = 0.61 \, R_{25}$, calculated for the $0°$ to $25°C$ range; the other is based on $R_S = 1.67 \, R_{25}$, calculated for the $0°$ to $70°C$ range. If V_S is 5V and R_{25} is $10k\Omega$, the respective nominal values of R_S would be $16.4k\Omega$ and $6.0k\Omega$. The useful ranges would be $-10°$ to $+30°C$ and $-5°$ to $+70°C$, with respective temperature sensitivities of $55mV/°C$ and $45mV/°C$.

A final example of modifying the transducer circuitry is the pair of circuits in Figure 5-9, where the output of a bridge preamplifier modulates the bridge's excitation. The simplified circuit of Figure 5-9a is shown in a practical embodiment, using the 2B31 bridge-signal conditioner, in Figure 5-9b. The adjustment permits a degree of over- or under-correction to partially compensate for the nonlinearity of the device (e.g., an RTD) as well as that of the bridge.

In Figure 5-9a, the output of the amplifier, E_O, is related to the fractional resistance deviation, X, and the excitation voltage, V_{IN}, by an expression of the form,

$$E_O = K \, V_{IN} \, f(X) \qquad (5.5)$$

$$* \quad \frac{d\left[\dfrac{V_{OUT}}{V_S}\right]}{dT} = \frac{-1.443 \times 10^{-6} \dfrac{R_{25}}{R_S} \dfrac{4016}{T} \, \epsilon^{4016/T}}{\left[1.443 \times 10^{-6} \dfrac{R_{25}}{R_S} \, \epsilon^{4016/T} + 1\right]^2} \qquad (5.4)$$

We feed back a fraction of E_O, βE_O, to make V_{IN} a function of E_O,

$$V_{IN} = V_{REF} + \beta E_O \tag{5.6}$$

Plugging 5.6 into 5.5, we find that

$$E_O = \frac{K V_{REF} f(X)}{1 - K \beta f(X)} \tag{5.7}$$

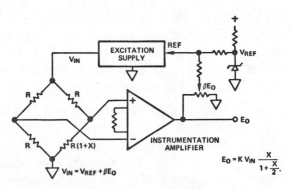

a. A small fraction, β, of the amplified bridge output is fed back to modulate the excitation voltage. β is adjusted to linearize the bridge output as a function of the deviation, X.

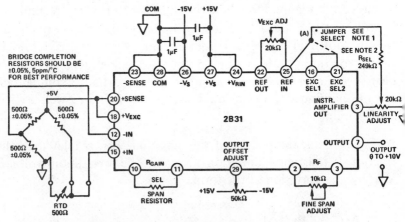

b. Practical linearizing circuit, employing the 2B31 signal conditioner

Figure 5-9. Transducer nonlinearity correction using feedback to affect bridge excitation.

If the bridge response is of the form, $f(X) = X/(1 + X/2)$,

$$E_O = \frac{K\,V_{REF}\,X}{\left(1 + \dfrac{X}{2}\right)\left(1 - K\beta\,\dfrac{X}{1+X/2}\right)}$$

$$= K\,V_{REF}\,X\,\frac{1}{1 + X(1/2 - K\beta)} \tag{5.8}$$

In order to cancel the nonlinearity, by making the denominator$= 1$,

$$K\beta = 1/2 \tag{5.9}$$

In Figure 5-9b, this principle is applied to a bridge circuit, in which a 2B31 signal conditioner provides excitation and amplification for an RTD measurement. The sense of the feedback is determined by whether the nonlinearity is concave upward or concave downward (jumper A to pin 21, or to pin 25). The magnitude of the correction is determined by the resistor, R_{SEL}, and the *linearity adjust* pot provides a fine trim.

If an RTD is to be used, the adjustment can be made efficiently, without actually changing the temperature, by simulating the RTD with a precision resistance decade. The offset is adjusted at the low end of the resistance range, the fine span is adjusted at about one third of the range, and the linearity is adjusted at a resistance corresponding to full-scale temperature. One or two iterations of the adjustments will probably be found necessary because of the interaction of linearity error and scale-factor error. This circuit's applications are not restricted to RTD's; it will work in most cases where bridges are used—e.g., load cells and pressure transducers.

Analog processing of the transducer's output signal is used (where necessary) to compensate for nonlinearity in the transducer, the associated circuitry (such as bridges), or both. Depending on the nature of the nonlinearity, circuits used for linearizing may include:

Devices having inherently complementary nonlinearity (for example, circuits having logarithmic response to compensate for exponential functions of the measurand—Figure 5-10)

Simple analog computing circuits that provide functions complementary to the known functional relationships of the transducer circuitry (for example, circuits involving analog multi-

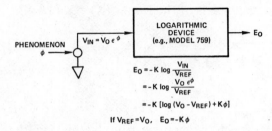

Figure 5-10. *Logarithmic device performs exponential and log operations. If V_{IN} is a log function of the phenomenon, the log device is connected for exponential operation. As shown here, for an exponential voltage (or current) input, the device—connected for log operation—will provide a linear output.*

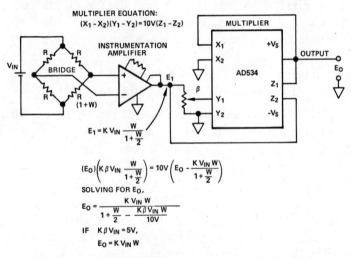

Figure 5-11. *Nonlinearity correction for large-deviation off-null bridge using a single AD534 analog multiplier. Adjust β to trim overall linearity.*

pliers* to compensate for off-null bridge nonlinearity—Figure 5-11)

Analog circuitry using analytic or piecewise-linear approximations to compensate for nonlinearities having an arbitrary form (for example, thermocouples and RTD's); both approaches are described comprehensively in the *Nonlinear Circuits Handbook*.[1]

---

*An analog multiplier, as shown, provides an output voltage that is equal to the true product of two input voltages, multiplied by a scale constant. A good introduction to analog multipliers may be obtained via a copy of the Analog Devices *Multiplier Application Guide* (1978), available upon request.

[1] Sheingold, D.H., Ed., *Nonlinear Circuits Handbook—Designing with analog function modules and IC's*, Analog Devices, Inc., 1974, pages 43-57 and 92-97

An example is worked through involving the linearization of a chromel-constantan thermocouple, for the range of temperature 0° to 650°C, to within ±1°C, with both smooth and piecewise-linear approximations. Though the argument is beyond the scope of this book, Figure 5-12 shows the nonlinear thermocouple response and the theoretical errors, Figure 5-13 shows the analytic approximation, using a Model 433 multifunction component to generate a $Y(Z/X)^{3.512}$ approximation; and Figure 5-14 shows a piecewise-linear (segmented) approximation using op amps as *ideal diodes*. Both circuits have been built, and they work.

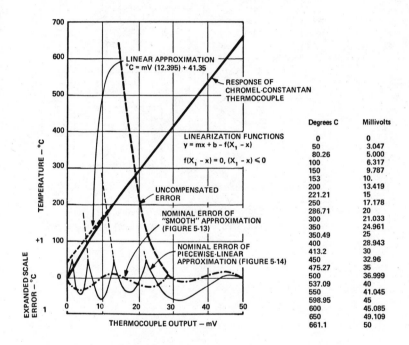

Figure 5-12. Nonlinear thermocouple response and theoretical residual errors, using two different linearizing functions. Read errors on expanded scale.

In this chapter, we have discussed the principles and a variety of techniques for offsetting and linearizing the outputs of transducers. In the next chapter, before proceeding to the Applications section, we will consider approaches to overall solutions to interfacing problems, expressed in terms of a couple of actual examples and discussed by a designer.

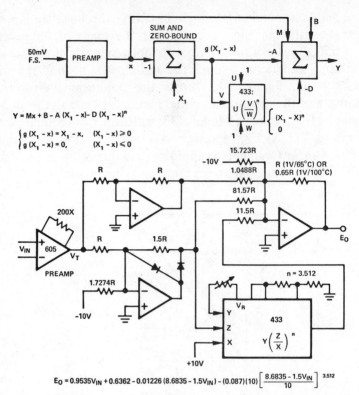

$$Y = Mx + B - A(X_1 - x) - D(X_1 - x)^n$$

$$\begin{cases} g(X_1 - x) = X_1 - x, & (X_1 - x) \geq 0 \\ g(X_1 - x) = 0, & (X_1 - x) \leq 0 \end{cases}$$

$$E_O = 0.9535 V_{IN} + 0.6362 - 0.01226(8.6835 - 1.5V_{IN}) - (0.087)(10)\left[\frac{8.6835 - 1.5V_{IN}}{10}\right]^{3.512}$$

Figure 5-13. Block diagram and circuit for linearizing, using smooth approximation. Theoretical error is shown in Figure 5-12.

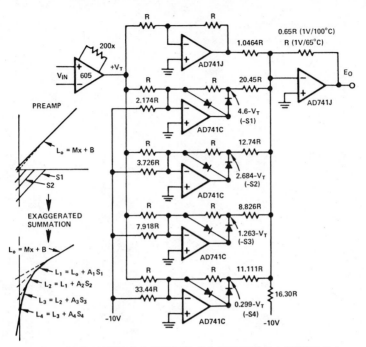

$$E_O = 0.9557 V_{IN} + 0.6135 - 0.0489(4.6 - V_{IN}) - 0.0785(2.684 - V_{IN}) - 0.1133(1.263 - V_{IN}) - 0.09(0.299 - V_{IN})$$

Figure 5-14. Circuit for linearizing using piecewise-linear approximation

Overall Considerations
2 Interface-Design Examples

Chapter 6

In the preceding chapters, we have considered separately the various elements of the transducer interfacing problem. In the chapters that follow, in Section II, a variety of applications are described in differing levels of detail. The focus of the Applications section is on circuitry employing lower levels of system integration. It is principally intended for those readers who are interested in design and are not considering purchase of a packaged interface, replete with interface cards, data acquisition, digital processing capability, software, and system support.*

As a transition to the Applications section, this chapter treats some of the questions that must be considered to obtain a good trouble-free design; it also brings together in a pair of design examples some of the factors discussed in the earlier chapters.

INTERFACE DESIGN QUESTIONS

When an interface problem is considered, there are some basic questions that must be answered before a level is reached that is relevant to the readers of this book:

What kind of physical information is really needed in electrical form to accurately reflect the state of the environment?

What is the best way to obtain the needed information?

Once these questions have been resolved, and a set of measurements and the transducers to implement them have been chosen, then the technical and economic issues relating to the choice of interface circuit/system can be considered. Some of the seminal technical questions (they will lead to additional questions) are:

*Such as MACSYM Measurement And Control (sub) SYsteMs.

What accuracy and stability are desired? Over what range of ambient (and measured) temperature?

What is the sensitivity of the transducer under consideration? As a matter of good design practice, it is desirable to have as high a sensitivity as the economics permit, and to give up as little as possible in attenuation. Obtaining high sensitivity may require sacrifices in linearity, dynamic range, and other characteristics, but it may be well worth it if resolution is to be preserved against noise and drift.

What are the energy (voltage or current) and impedance levels at the output of the transducer, and what are the implications for circuit wiring and performance? For example, low-level transducers, like thermocouples, require good low-noise, low-drift amplifiers to stably resolve small temperature changes; high-impedance devices, such as piezoelectric transducers, require attention to wiring capacitance and amplifier input characteristics (input impedance, leakage current, *mode*: charge or voltage).

What are the error sources in the transducer? Besides such obvious considerations as noise and linearity, what are the effects of temperature on a load cell or pressure transducer; what are the effects of physical forces on an RTD? Is electrical interference characterizable? What kinds of correction capability must the interface have?

Is the output signal single-ended or differential? Is there any choice? What is the common-mode level; what are the sources of common-mode error? Will an instrumentation amplifier or an isolator be necessary to remove ground noise and 60Hz artifacts?

What sources of noise and interference are likely to be encountered? How much? To what degree will judicious wiring and grounding practices help?

What is the fastest rate-of-change of transducer output signals? What are the tradeoffs affecting choice of filtering circuitry—amplitude and phase errors vs. noise and interference? Check for sufficient bandwidth—both for signal and for common-mode rejection—in the electronic circuitry.

For how long and over what range of temperature excursion must amplifiers and associated components function, and with what degree of repeatability and stability? This important consideration directly influences the certainty and credibility of the data at the output of the interface.

Is offsetting required? Linearizing? If they are, should they be implemented in the transducer, in the preamplifier or subsequent analog stages, or in digital form?

Does the transducer require some form of excitation? If so, what are the power, accuracy, and stability requirements of the excitation source? Must it have absolute stability, or is some form of ratiometric measurement more economically feasible.

Attention to these criteria, and to others which logically follow from them in any given situation, is the key to success in almost any interface job. The successful interface designer is typically an *optimistic skeptic*–confident of success and wary of the forces of nature. Even a seemingly mild requirement should call for careful thought to avoid surprises.

THERMOMETER EXAMPLE

In this application, there is a need to measure temperatures from 0°C to +100°C, to within 1.0°C, at low cost, at a remote location several hundred feet from the instrumentation. The ambient temperature in the vicinity of the instrumentation is expected to be 25°C ±15°. A number of possible transducers will operate over the specified range, but the requirement for a remote measurement suggests the use of the current-output two-wire AD590 semiconductor temperature sensor, because the current is unaffected by voltage drops and induced voltages.

Consulting the "Accuracies of the AD590" (see the Appendix), we find that the AD590J, with two external trims, would be suitable; its maximum error over the 0°C to 100°C range is 0.3°. This permits an allowance of 0.7° for all other errors. If a tighter tolerance were required, it would be worthwhile to consider using the AD590M with two trims, for an error *below 0.05°C.*

Since AD590 measures absolute temperature (its nominal output is $1\mu A/K$), the output must be offset by $273.2\mu A$ in order to read out in degrees Celsius. The output of the AD590 flows through a $1k\Omega$ resistance, developing a voltage of $1mV/K$ (Figure 6-1). The output of an AD580 2.5-volt reference is divided down by resistors to provide a $273.2mV$ offset, which is subtracted from the voltage across the $1k\Omega$ resistor by an AD521 instrumentation amplifier. The AD521 provides a gain of 10.0, so that the output range, corresponding to 0° to 100°C, is 0 to 1.00V ($10mV/°C$).

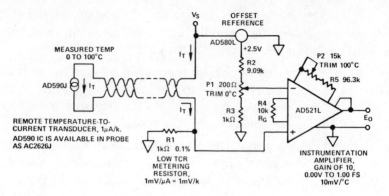

Figure 6-1. Thermometer circuit

The desired system accuracy is to within 1.0°C; as noted, all errors other than that of the AD590 must contribute the equivalent of less than 0.7°. It will be helpful to assemble an error budget for the circuit, assessing the contributions of each of the elements (Table 1). Errors will be expressed in degrees Celsius.

AD590 regulation. If the AD590 is excited by a voltage source of between 5 and 10V, the typical regulation is 0.2µA/V (0.2°C/V). With 1% source regulation, this contribution will be

0.01°C

AD590 linearity error. Total error for the AD590J, over the 0° to 100°C range, with two trims, is 0.3°C. Those trims will be the gain and offset trims for the whole circuit, accounting for resistor and ratio errors, AD521L gain, offset and bias-current errors, AD580L voltage error, and the AD590J's calibration error.

0.3°C

R1 temperature coefficient. Since R1 is responsible for the conversion of the AD590's current to voltage, high absolute accuracy is important. Consequently, we would expect to use a device having 10ppm/°C or less in this spot. For ±15°C, the maximum error is $373.2\mu A \times 10^{-5}/°C \times 15° = 0.06\mu A$

0.06°C

AD580 temperature coefficient. The specified tempco for the AD580L is 25ppm/°C typical (61ppm/°C max over the range 0° to 70°C). Since operation is over a narrow range, the typical

TABLE 1. DEVICE SPECIFICATIONS PERTINENT TO THE ANALYSIS IN THE TEXT

(typical at 25°C and rated supply voltage unless noted otherwise)

Parameter	Condition	Specification
AD580L 2.5V VOLTAGE REFERENCE		
Output voltage	V_S = +15V	2.450V min, 2.550V max
Input voltage, operating		30V max, 7V min
Line regulation	7V ≤ V_{IN} ≤ 30V	2mV max
Temperature sensitivity	0 to 70°C	4.3mV max, 25ppm/°C, typ
Noise	0.1 to 10Hz	60μV, p-p
Stability (drift with time)	long term	250μV (0.01%)
	per month	25μV (10ppm)
AD590J 1μV/K TEMPERATURE TRANSDUCER		
Output current	Nominal at 25°C (298.2k)	298.2μA
Input voltage, operating		30V max, 4V min
Calibration error	25°C, V_S = 5V	±5°C max
Linearity error	Two trims, 0 to 100°C range	0.3°C max
Repeatability	per month	0.1°C max
Long-term drift		0.1°C max
Noise spectral density		40pA/$\sqrt{Hz}$
Power-supply rejection	+5V ≤ V_S ≤ +15V	0.2μA/V
Operating range		-55°C to +150°C
AD521L DIFFERENTIAL INSTRUMENTATION AMPLIFIER		
Gain equation (volts/volt)	Nominal	$G = R_S/R_G$
Error from equation	Untrimmed	(±0.25 – 0.004G)%
Nonlinearity	±9V output	0.1% max
Gain tempco	0 to 70°C	±(3 ± 0.05G)ppm/°C
Voltage offset	Input	1.0mV max
	Output	100mV max
Voltage offset tempco	Input, 0 to 70°C	2μV/°C max
	Output, 0 to 70°C	75μV/°C max
Voltage offset vs. supply	Input	3μV/%
	Output, untrimmed*	0.5mV/%
Bias current	25°C	40nA max
Bias current tempco	0 to 70°C	500pA/°C
Input impedance	Common-mode	6 × 10^{10}Ω‖3.0pF
Common-mode rejection	G = 10, dc to 60Hz, 1kΩ source imbalance	94dB min
Voltage noise	G = 10, 0.1Hz to 10Hz, p-p, RTO	225μV

*Can be reduced by trimming the output offset.

value is most useful, unless the AD580 has a critical effect on the overall error. 25×10^{-6}/°C × 273mV × 15° = 0.1mV.

0.10°C

Resistive divider tempco. The absolute values of R2 and R3 are of considerably less importance than their ability to track. 10ppm/°C is a reasonable value for tracking tempco. 10^{-5}/°C × 273mV × 15° = 0.04mV

0.04°C

Common-mode error. At a gain of 10, the minimum common-mode error of the AD521L amplifier is 94dB, one part in 50,000 of the common-mode voltage (273mV), or 5μV (negligible)   0.0°C

AD521 temperature coefficient. The specified input offset tempco for the AD521L is 2μV/°C max, and the output offset tempco is 75μV/°C max (7.5μV/°C, referred to the input), for a total of 9.5μV/°C R.T.I. 9.5μV/°C × 15° = 143μV.   0.14°C

AD521 bias-current tempco. The maximum bias-current change is 500pA/°C × 30° (range) = 15nA. The equivalent offset-voltage change is 15nA × 1kΩ = 15μV.   0.02°C

AD521 gain tempco. The circuit will be calibrated for correct output at 100°C by trimming of the gain of the AD521 at a 25°C ambient temperature. Variation of gain will cause output errors. The specified gain tempco at a gain of 10 for the AD521L is 3.5ppm/°C typical. If max is arbitrarily assumed to be ten times worse, and the resistors contribute 15ppm/°C additional, the maximum error will be $50 \times 10^{-6}/°C \times 100° \times 15° = 0.075°$ 0.075°C

AD521 nonlinearity. The 0.1% nonlinearity specification applies for a ±9V output swing; for a 1V full-scale swing, it may be reasonable to expect a tenfold improvement, or a 1mV linearity error, equivalent to 0.1°C   0.1°C

Total error (worst case)   0.84°C

This means that, once the circuit has been calibrated at 0°C and 100°C (25°C ambient), the maximum error at any combination of measured and ambient temperatures can reasonably be expected to be less than 1°C.

If the summation were root-sum-of-squares, instead of worst-case, the error would come to less than 0.4°. This suggests that the design is quite conservative, since the probability of worst-case error is low; also (with some risk), it suggests that if an AD590M were used in the same design, temperature could be measured to within 0.25°C over the range. Naturally, every precaution should be taken to avoid additional errors attributable to either Murphy's or Natural Law. Aside from errors attributable to ambient temperature variations, this simple interface will require some form of protection from extraneous signals. Shielding and grounding should follow the practice suggested earlier in this book. In addition, capacitance across R1 will help reduce the effects of any ac currents induced in the twisted pair. Power supplies must be chosen to minimize error due to sensitivity of any of the elements to power-supply voltage changes, and bypassed to minimize coupling of interference through the power-supply leads.

Although this is one of the simplest forms of interface, it is impressive to consider the preparation required to ensure performance to within the desired specs. As the level of accuracy increases, additional parameters and issues could be mentioned, but the point that is made would begin to suffer from overkill. As the statement of a measurement problem becomes more complex and demanding, the increased scope demands more experience and subtlety of thought on the part of the designer.

EXAMPLE: HIGH-RESOLUTION SCALE

A high-sensitivity nutritional experiment at the Massachusetts Institute of Technology required an instrument that would repeatably resolve 0.01 pound, out of 300 pounds full-scale, about 33 parts per million.* In addition, the instrument was required to be hand-carryable, be free from adjustments ("forever"), and have an absolute accuracy to within 0.05lb.

Such instruments are based on the availability of pre-trimmed fully calibrated load-cell platforms having low temperature drift, excellent linearity, and compensation for off-center loading. In this case, a BLH Electronics PL-250 platform was chosen (Figure 6-2). Because the PL-250 has an output of 1.5 millivolts per volt of excitation at 250 pounds, the output produced by a 300-lb load, with 10 volts of excitation, is 18 millivolts (full scale!)

The resolution corresponding to 18mV full-scale is 1/30,000 of 18 millivolts, or 600 *nanovolts*! While the specified clinical temperature range of ±5° is usually benign in terms of error production, it is definitely a factor in this application, since the bridge readout

---

*One small doughnut bite for a 300-lb subject, or a shot-glass of whiskey in a full-size 1977 Cadillac.

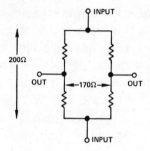

a. Simplified schematic

	GUARANTEED	AS MEASURED OR USED
ABSOLUTE MAXIMUM EXCITATION	15V	10V
CALIBRATION ACCURACY	0.25%	0.11%
NON-LINEARITY	0.05%	0.015%
REPEATIBILITY	0.02%	0.01%
TEMPERATURE COEFFICIENT	25ppm/°C	1.1ppm/°C
RANGE	0–300 Lbs.	20–30°C

b. Pertinent specs

Figure 6-2. Data on the BLH–PL250 platform used in this application

system must have less than 120nV/°C drift, and in fact, all drifts (including drift with time) must amount to less than 600nV.

In order to perform the differential measurement, the first kind of amplifier considered is the differential instrumentation amplifier. However, none is known to be available commercially with adequate drift performance. An additional problem is the need for high common-mode rejection: in order for common-mode error to be negligible, the CMR must be upwards toward 140dB, if a one-sided 10V excitation supply is used. A split supply (±5V) will ease this aspect of the problem, but the problem of drift remains.

The most economical solution calls for a chopper amplifier, because chopper amplifiers can be obtained with drifts less than 100nV/°C. Although single-ended chopper amplifiers, such as Model 261K, cannot normally take a differential measurement, the output of a bridge can be amplified as a single-ended signal if the *excitation supply* can float. This also eliminates the common-mode problem, since a truly floating source has no common-mode potential. So this approach yields the desired low drift and solves the common-mode problem (Figure 6-3).

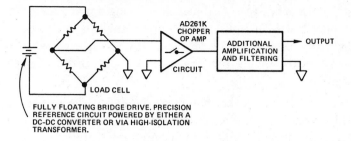

FULLY FLOATING BRIDGE DRIVE. PRECISION
REFERENCE CIRCUIT POWERED BY EITHER A
DC-DC CONVERTER OR VIA HIGH-ISOLATION
TRANSFORMER.

Figure 6-3. Block diagram of high-precision scale

However, there are still some difficulties. Besides the possibility that a floating supply may invite line-frequency pickup, there is the important consideration that the floating supply must furnish a regulated voltage *at least as stable as the 33ppm instrument specification*. This degree of stability is necessary because, with the bridge floating, a ratiometric measurement is no longer practical. If the bridge were a null type (forced to null by a servo loop), the precision of the excitation source would not cause difficulty; however, with the bridge operating off null, the output voltage depends directly on the excitation voltage.

The first difficulty, line-frequency pickup, can be of minor importance, because the impedance of the excitation source to ground is only about 100Ω. This means that, with good shielding, the combination of low coupling capacitance and low shunt resistance to ground will result in negligible pickup. Also, the isolation characteristics of the floating source at dc need not be outstanding (30,000 × 100Ω = 3MΩ).

The only important difficulty with this approach, then, is the requirement for a separate high-stability voltage drive for the bridge. While not trivial, a stable, floating bridge drive can be implemented with a power amplifier, driven by a high-stability reference, such as the AD581L or AD2700, powered by an isolated dc-to-dc converter, such as those in the 940 series (or via a high-isolation line transformer). The block diagram of Figure 6-3 illustrates this concept. Design details appear in the Applications section (Figure 12-5).

CONCLUSION

In this chapter, we have discussed the basic ingredients of successful application; and we have considered two examples that illustrate approaches to transducer interfacing applications. The first example shows an approach to error analysis of a simple and straightforward interfacing problem. The second example considers approaches to solving the key problems of a difficult interfacing assignment.

This chapter concludes what might be called the tutorial section of this book. The following chapters discuss some fifty real-world interfacing configurations. They begin immediately.

Applications

APPLICATIONS

The applications described in the following pages vary in degree of detail, from circuits that demonstrate techniques to detailed descriptions of measurement and control circuitry employing transducers. Many of the circuits are simple and straightforward and require little comment; others merit detailed discussion and explanation. Most (if not all) have been used in real-world applications.

The organization of Part Two is roughly parallel to that of the descriptions of transducers in Chapter 1; transducer applications are arrayed by phenomenon, by measurement technique, and by level of detail and manner of readout.

In general, to avoid excessive redundancy, we have not repeated circuits in one medium where they apply to commonly employed transducers in other media; for example, the discussion of thermal measurements to measure other quantities is limited in scope, as is the use of force and pressure to measure level and flow.

The final chapter of this section is a collection of miscellaneous circuits, most of them pertaining to analog signal processing. Though they could have been incorporated in specific applications, they are of sufficiently general applicability to warrant separate treatment.

Whether you are looking for a specific circuit or browsing for ideas, you should find the circuits in this section interesting.

Thermoswitches
and Thermocouples

Chapter 7

THERMOSWITCHES

Thermally sensitive switches are easily interfaced in a variety of applications calling for simple and reliable circuitry. Figure 7-1 shows two interface circuits involving thermal switches. In (a), a dual-contact mercury thermoswitch is used on an assembly line as an inexpensive check of temperature in a small component oven. The thermometer bulb is inserted into the oven. If the temperature is within the desired limits, the lamp is lit.

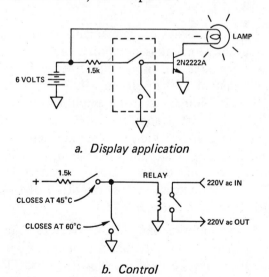

a. Display application

b. Control

Figure 7-1. Thermoswitch applications (see also Figure 1-3)

In (b), two thermoswitches monitor the wall temperature of a chemical vat, which must remain between 45° and 60°C, while it is being filled. If the temperature is between those limits, the relay is energized, permitting the vat to be filled; temperatures outside the limits de-energize the relay, stopping the filling process. The 1.5kΩ resistors protect the relay contacts from passing excessive current (*snap-disc* relays have specifications on minimum (dry-circuit) current, as well as maximum contact current).

AMBIENT-REFERENCED THERMOCOUPLES

As we have noted in Chapter 1, thermocouples require cold-junction compensation if they must resolve temperature changes with precision better than the ambient temperature range at the cold junction. However, for high-temperature measurements to within a few percent, the cold junction may often be profitably left at room ambient.

Suppose, for example, that a Type S thermocouple* is used to measure temperatures of the order of 1500°C within a furnace, and the ambient temperature of the cold junction is 25°C ±15°C. Since the sensitivity of the thermocouple is 12μV/°C at 1500°C, and a change from 10° to 40°C at the cold junction produces a change of 180μV in the net output voltage, the equivalent ΔT at the active junction is 15°C for a full-scale change at the cold junction, or 1% of 1500°C.

Figure 7-2 shows two ways of implementing this application. In (a), where the environment is not noisy and leads are short (or battery power can be used), an op amp provides the amplification. The indicated gain of 500 results in a full-scale output between 7.5V and 8.0V at 1500°C, with the amplifier zeroed. The amplifier's 3μV/°C max drift is somewhat less than that of the cold junction, so the overall precision can be expected to be to within about 2% at 1500°C. If calibration is desired, the amplifier's zero can be adjusted to take the cold-junction voltage into account, with an insignificant increase in drift (an offset of 142μV causes a change in drift rate of about 0.5μV/°C).

In (b), an instrumentation amplifier is used to reject common-mode noise. If there is no conductive return path from the thermocouple, resistance may be used (as shown) to provide a path for the amplifier's bias currents. The gain is determined by the ratio of the *gain* and *scale* resistors. The maximum offset drift spec of

---

*Abbreviated tables of thermocouple voltage and sensitivity for all popular types will be found at the end of this chapter.

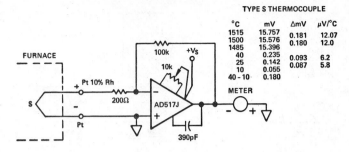

TYPE S THERMOCOUPLE

°C	mV	ΔmV	μV/°C
1515	15.757	0.181	12.07
1500	15.576	0.180	12.0
1485	15.396		
40	0.235	0.093	6.2
25	0.142	0.087	5.8
10	0.055		
40 – 10	0.180		

a. Operational amplifier

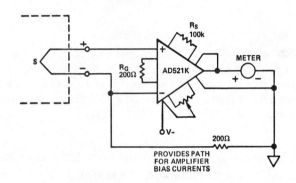

b. Instrumentation amplifier

Figure 7-2. Ambient-referenced thermocouple applications

the AD521K at high gains is comparable to that of the cold junction.

COLD-JUNCTION COMPENSATION

If ambient temperature variation of the cold junction can cause significant error in the output of a thermocouple pair, there are two alternatives: maintain the cold junction at constant temperature, by some such technique as an ice bath or a thermostatically controlled oven, or subtract a voltage that is equal to the voltage developed across the cold junction at any temperature in the expected ambient range.

The latter option is usually the easier to implement. Figure 7-3, which is similar to Figure 5-3, shows a simple application, in which the variation of the cold-junction voltage of a Type J thermocouple—iron(+)—constantan—is compensated for by a voltage developed in series by the temperature-sensitive output current of an AD590 semiconductor temperature sensor.

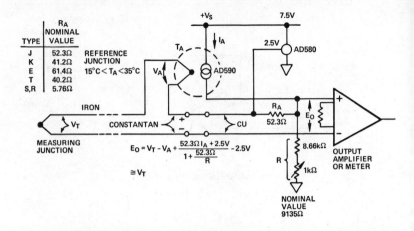

Figure 7-3. Cold-junction compensation circuit for type J thermocouple, employing AD590 to sense cold-junction ambient temperature and provide compensation over the ambient range 15°C to 35°C. Voltage across R_A compensates for V_A, and the 2.5V reference provides a 273µA current via R to offset the AD590's output current at 0°C.

The circuit is calibrated by adjusting R_T for proper output voltage with the measuring junction at a known reference temperature and the circuit near 25°C. If resistors with low tempcos are used, compensation accuracy will be to within ±0.5°C, for temperatures between +15°C and +35°C. Other thermocouple types may be accommodated with the standard resistance values shown in the table. For other ranges of ambient temperature, the equation in Figure 7-3 may be solved for the optimum values of R_T and R_A. If an instrumentation amplifier is used, gain and offset specifications should be appropriate for the temperatures being measured, the required precision, and the sensitivity of the thermocouples employed.

THERMOCOUPLE-BASED TEMPERATURE CONTROL

In the circuit of Figure 7-4, a thermocouple measures an object's temperature in an oven. The measured value is compared with a setpoint, and a heater is operated when the temperature drops below the set value, for temperatures to beyond 300°C.

The thermocouple is Type T—copper(+)–Constantan; the temperature-sensitive current of an AD590, flowing through the 40.2Ω resistance, provides a 40µV/°C cold-junction-compensation voltage. The AD590 is kept in intimate thermal contact with the cold

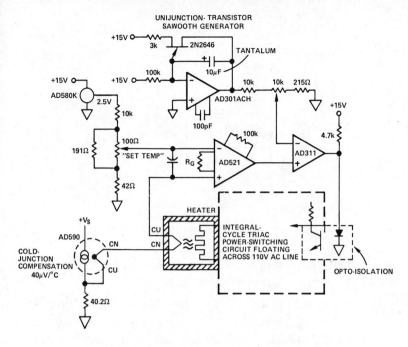

Figure 7-4. Temperature control circuit

junction. The difference between the set-point voltage and the compensated thermocouple output (minus about 11mV due to the AD590's 273.2μA output at 0°C) is amplified by the AD521, which is set for a gain of about 100.

The heater is ac-operated via a Triac, which provides current in increments of whole cycles of line frequency. The number of cycles is determined by comparing the output of the AD521 and a 1Hz negative-going sawtooth in the AD311 comparator. The larger the error (setpoint minus measured temperature), the greater the number of cycles of power to the heater per second. The sawtooth is generated by an integrator that is periodically reset by a unijunction transistor.

This scheme provides sensitive, fast-responding, and essentially smooth control of temperature at the measuring thermocouple.

ISOLATED THERMOCOUPLE MEASUREMENT

In Figure 7-5, the small size of a surgically implanted thermocouple and the safety provided by an isolation amplifier combine to provide safe and accurate monitoring of the temperature in a labora-

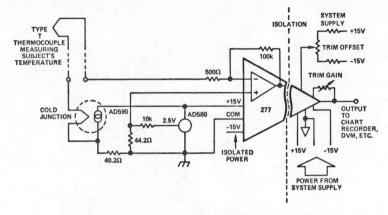

Figure 7-5. Isolated thermocouple measurement

tory animal's cerebral cortex. This circuit was used to study body-temperature regulation in sleeping monkeys.

Isolation is provided by the low-drift Model 277, which has an uncommitted operational amplifier for a front end and a low-impedance operational-amplifier output. It also provides auxiliary isolated front-end power, which is useful for driving the cold-junction compensation circuitry.

The Type T thermocouple is compensated by a cold-junction circuit similar to that of Figure 7-4, using an AD590's output current to develop a series voltage drop that matches the cold junction's $-40\mu V/^{\circ}C$ tempco. An AD580 and a voltage divider provide the 11mV of fixed offset required to null out the net output of the AD590 at $0^{\circ}C$. (In the previous application, this function was served by the set-point adjustment.)

The net thermocouple output is amplified (X200), transmitted across the amplifier's isolation barrier, and amplified further in the output amplifier. Any necessary calibration or offset adjustments are employed in conjunction with the output stage.

The animal is fully protected from shock hazard due to grounding problems because there is no galvanic (and little capacitive) path from its body to ground.

At the time this circuit was designed, the 2B50 isolated thermocouple amplifier was not available. If it were, the design problem would have been further simplified, since much of the external circuitry provided by the user would not be required.

THERMOCOUPLE-TO-FREQUENCY CONVERSION

If an analog quantity is converted to frequency, it can easily be isolated by optical techniques; and it can be converted to digital by counting over a predetermined interval. The AD537 V/f converter is a useful tool for performing such conversions, because it is easy to apply, requires little power, and is low in cost.

As Figure 7-6 shows, the heart of the AD537 is a current-to-frequency converter; frequency is determined by an externally connected capacitance. The current that establishes the frequency is buffered by a differential input stage that works in much the same way as an op amp. Thus, if a positive voltage, V_{IN}, is applied at the + input, and a resistance, R, is connected between the negative input and ground, the current that flows in the buffer output is V_{IN}/R. If a resistance, R_L, is connected between a voltage source, V_R, and the negative input terminal, it will add an offset current, $-V_R/R_L$, when V_{IN} is zero. (The effective resistance, that determines the sensitivity to V_{IN}, will then depend on the parallel combination of R and R_L). The frequency output is provided via an open-collector output stage, which can be referenced to an arbitrary voltage level. Frequency is nominally given by

$$f = 0.1 \frac{I}{C} \text{ hertz} \qquad (7.1)$$

where I is in amperes, C is in farads, and 0.1 has the required dimensions, of HzF/A; for example, if $C = 0.001\mu F$ and $I_{max} = 1mA$, then $f_{max} = 100kHz$.

The AD537 also has a reference output that is nominally equal to 1V, and a temperature-sensitive output of 1mV/K. The use of the reference output is demonstrated in this application; the use of the temperature-sensitive voltage in direct temperature-to-frequency conversion is discussed in Chapter 10.

In the application considered here, a Type E thermocouple—nickel–10% chromium(+)–Constantan—is used to measure temperatures in the range 700°C to 400°C. In this range, the thermocouple is quite linear, with an average sensitivity of 80.6μV/°C, and a full-scale output of 53.11mV at 700°C. We wish to have a frequency output of 10Hz/°C (7kHz full-scale). If precise operation at temperatures down to 0°C is imperative, some sort of linearizing would be necessary (see the Analog Devices *Nonlinear Circuits Handbook*, pages 92–97), but in many cases, such as the one treated here, operation is needed over only part of the range.

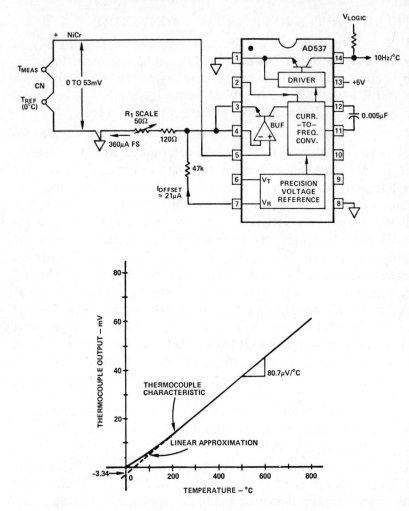

Figure 7-6. Type E thermocouple interface with the AD537 VFC (DIP package).

The circuit shown in Figure 7-6 provides good accuracy from +300°C to +700°C. If the temperature-voltage curve is extrapolated to 0°C, an offset of –3.34mV is seen to be required for the best fit. The small amount of current corresponding to this voltage is introduced without an additional calibration step by using the +1.00V output of the AD537. To adjust the scale, the thermocouple should be exposed to a known reference temperature near the upper end of the scale, and the frequency should be adjusted to the corresponding value with R1. The error should be below 1° over the range 400°C to 700°C.

THERMOCOUPLE-to-4-20mA TEMPERATURE TRANSMITTERS

The 2B52 and 2B53 families of signal-conditiong modules provide cold-junction compensation and amplification for all the popular thermocouple types. The output is in the form of current; a 4-to-20mA range corresponds to the total temperature span of interest. These devices are designed for two-wire operation; since they derive their power from the loop, both power and signals are transmitted over the same two wires. The 2B52 provides complete galvanic isolation between the thermocouple and the current loop; the 2B53, for less-demanding applications, has connections in common between input and output.

Figure 7-7 shows how a 2B52 is typically connected. Screw terminals are provided for connections to the external system, and the span and offset can be adjusted within the protective enclosure. The device provides a system solution to the problem of reliably measuring temperature and transmitting the information in a standard analog format, with a minimum of interface-design decisions.

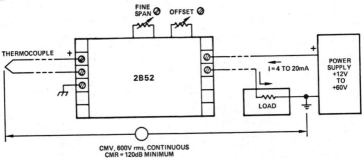

Figure 7-7 2B52 two-wire thermocouple signal transmitter. Gain and offset settings are determined by the temperature range. Open thermocouple is indicated by upscale saturation (downscale optional).

ISOLATED MULTIPLEXING OF THERMOCOUPLES

For applications where a number of thermocouples (similar or different) must be read out, a ready all-electronic system solution, competitive with flying-capacitor techniques, is available in the form of the 2B54 four-channel low-level isolator and the 2B56 cold-junction compensator (see also Figure 15-14).

The 2B54's four input channels are filtered and galvanically isolated from one another—and from the output—and protected for 750V rms common-mode or interchannel voltage, as well as 130V rms ac differential input voltage. Individually adjustable amplification (ranging from 25V/V to 1000V/V, for input spans of ±5mV to ±200mV and output span of ±5V) is provided for each channel, with low drift and noise. The 2B54 can detect open inputs, and the output is protected against continuous shorts to either supply or ground.*

If more than four channels are to be used, a number of 2B54's may be employed; a "three-state analog" output connection permits the outputs to be connected in common and enabled individually by a digital logic signal. Synchronized isolator drive circuitry eliminates beat-frequency errors.

Figure 7-8 shows an eight-channel temperature-measurement system employing eight thermocouples, two 2B54's and a 2B56 cold-junction compensator. The 2B56 is designed to operate with an external temperature-sensitive semiconductor element thermally integrated with the cold junction. The cold junction of each thermocouple will be compensated to 0°C to within ±0.8°C max over a range of ambient temperatures from +5°C to +45°C.

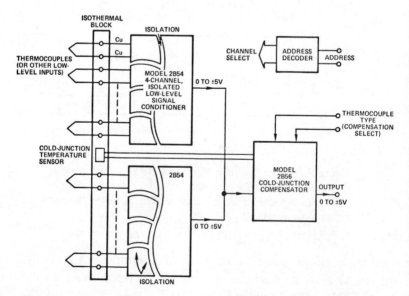

Figure 7-8. Temperature measurement system using isolated low-level multiplexer and cold-junction compensator

---

*The 2B54 can, of course, be used for isolating any combination of low-level sources (RTD's, strain gages, etc.); it is shown here with the thermocouples because of its affinity for the 2B56 cold-junction compensator.

As Figure 7-9 shows, the 2B56 has four digitally selectable compensation circuits, for J, K, T, and a user-determined type (or *none**). It is easy to see that a wide range of temperatures can be measured, using the optimum thermocouple for each range, and digital selection of compensation appropriate to the thermocouple type for each channel, as it is selected.

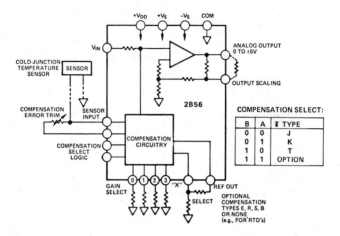

Figure 7-9. Functional Block Diagram of Cold-Junction Compensator

SCANNING DIGITAL READOUT AND PRINTING

The amplified thermocouple outputs from a single channel can, of course, be read out with a digital voltmeter.

With the AD2036 scanning thermocouple thermometer, 6 channels may be read out in °C or °F in a variety of ways (sequentially, single-channel continuously, or in a random sequence determined by BCD logic). The AD2036, operating with line voltage, provides isolation (for the six channels in common), cold-junction compensation, linearization, a/d conversion, and a variety of readout-control options. Three optional models will handle Types J, K, or T thermocouples, and nonisolated devices are available for operation with +5V or +12V power.

For a permanent record, without an expensive data logger, the AD2036 can be used with a number of available printers; for low-cost strip chart recording, an analog output is available. Figure 7-10 shows the details of an interface between the AD2036 and a Gulton ANP-9 thermal-head alphanumeric printer.

*e.g., if some of the multiplexed channels are RTD's or strain gages or thermocouples not requiring cold-junction compensation.

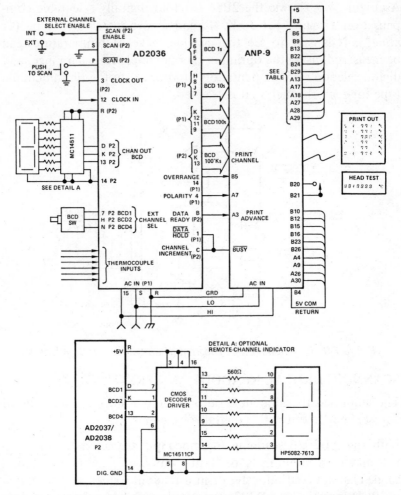

Figure 7-10. Interconnection of AD2036 and Gulton ANP-9 Printer

A scan of the channels is initiated via a pushbutton or other pulse source. When *data ready* goes high, $\overline{busy}$, from the printer, goes low. This *holds* the Data and Channel-Number outputs until the printer's $\overline{busy}$ goes high, releasing the *hold* and incrementing the channel counter. After 3.2s (in the standard unit), the *data ready* goes high and initiates a new cycle. After six cycles have been completed, the scan stops. Each time the scan is initiated, automatically or manually, the sequence will repeat.

For continuous scanning of all six channels, $\overline{scan}$ is held at logic zero. For continuous printing of a *single channel*, $\overline{scan}$ is held at logic zero, and the desired channel is selected, either by a manual switch setting (on the AD2036) or by external logic.

For external channel selection, the *scanner enable* line should be held low. Under external BCD control, the channel is immediately selected. If the $\overline{scan}$ line is pulsed low, the printer will print the selected channel's data six times and stop. If $\overline{scan}$ is held low, a continuous printout of the selected channel will result. Power (at +5V) is provided for external logic at the rear connector of the AD2036.

As the inset shows, the channel number (which is automatically printed as each channel's data is printed) may be *displayed* on a 0.3″ high-efficiency common-cathode H-P display. The channel BCD output operates a seven-segment decoder-driver, which, in turn, drives the LED. Power is furnished by the AD2036.

The detailed pin-to-pin connections between the AD2036 and the printer are listed in an application note available from Analog Devices.[1]

THERMOCOUPLE OUTPUT TABLES

Table 7-1 is an abbreviated listing of the most popular thermocouple pairs, condensed from data given in American National Standard C96.2-1973, published as ASTM document E 230-72. For the sake of compactness, thermocouple outputs, in millivolts, are given at each 100°C, along with the temperature coefficient at that temperature, in $\mu V/°C$. This form of presentation has the following virtues:

1. The basic data give a feeling for range and sensitivity of each thermocouple type. For example, one can quickly see that, although a Type E thermocouple is the most sensitive of all the types, it cannot be used for measurements at temperatures exceeding 1000°C.

2. In the absence of more-complete tables, it is easy to interpolate between the given values, to either a first order, using linear interpolation, or to better accuracy, using the variation of tempco between data points. For example, simple linear interpolation permits one to conclude that the output of a Type J thermocouple at 530°C is equal to 27.388 + 0.3(33.096 − 27.388) millivolts, or 29.100mV. The actual output, according to the above source, is 29.075mV. *(text continues on page 134)*

[1] "Interfacing the AD2036 6-Channel Digital Scanning Thermometer with the ANP-9 Thermal Printer," by Steve Castelli

THERMOCOUPLE RESPONSES

TABLE 7-1. VOLTAGE AS A FUNCTION OF TEMPERATURE

	B		E		J		K
Temperature °C	Output mV	Tempco μV/°C	Output mV	Tempco μV/°C	Output mV	Tempco μV/°C	Output mV
- 200	–	–	- 8.824	25.1	- 7.890	21.8	- 5.891
- 100	–	–	- 5.237	45.1	- 4.632	41.1	- 3.553
0	+ 0.000	–	0.000	58.7	0.000	50.4	0.000
+ 25	- 0.002	0.1	1.495	60.9	1.277	51.7	1.000
+ 100	+ 0.033	0.9	6.317	67.5	5.268	54.4	4.095
+ 200	+ 0.178	2.0	13.419	74.0	10.777	55.5	8.137
+ 300	0.431	3.05	21.033	77.9	16.325	55.4	12.207
+ 400	0.786	3.95	28.943	80.0	21.846	55.2	16.395
+ 500	1.241	5.0	36.999	80.8	27.388	55.9	20.640
+ 600	1.791	5.95	45.085	80.7	33.096	58.5	24.902
+ 700	2.430	6.8	53.110	79.8	39.130	62.3	29.128
+ 800	3.154	7.65	61.022	78.4	45.498	64.6	33.277
+ 900	3.957	8.4	68.783	76.7	51.875	62.4	37.325
+1000	4.833	9.1	76.358	75.0	57.942	59.2	41.269
+1100	5.777	9.75	–		63.777	57.8	45.108
+1200	6.783	10.35	–		69.536	57.2	48.828
+1300	7.845	10.9	–		–		52.398
+1400	8.952	11.2	–		–		–
+1500	10.094	11.6	–		–		–
+1600	11.257	11.7					
+1700	12.426	11.7					
+1800	13.585	11.5					

TABLE 7-2. TEMPERATURE AS A FUNCTION OF VOLTAGE READING

	B		E		J		K
mV	°C	°C/mV	°C	°C/mV	°C	°C/mV	°C
-10.000	–		–		–		–
- 5.000	–		- 94.4	21.70	- 109.1	25.10	- 153.7
- 2.000	–		- 35.3	18.40	- 40.8	21.10	- 53.1
- 1.000	–		- 17.3	17.60	- 20.1	20.40	- 25.9
0.000	+ 42.0	4°/μV	0.0	17.06	0.0	19.84	0.0
+ 1.000	449.6	220	16.8	16.64	19.6	19.25	+ 25.0
+ 2.000	634.2	160	33.2	16.21	38.9	19.08	+ 49.5
+ 5.000	1018.2	109	80.3	15.19	95.1	18.43	+ 122.0
+10.000	1491.8	87	153.0	14.02	186.0	18.03	246.3
+20.000	–		286.7	12.90	366.5	18.13	485.0
+30.000	–		413.2	12.47	546.3	17.57	720.8
+40.000	–		537.1	12.36	713.9	15.94	967.5
+50.000	–		661.1	12.47	870.2	15.79	1232.3
+60.000	–		787.0	12.71	1035.0	16.95	
+70.000	–		915.9	13.09	–		

∨—— R ——∨—— S ——∨—— T ——∨

Tempco μV/°C	Output mV	Tempco μV/°C	Output mV	Tempco μV/°C	Output mV	Tempco μV/°C	Temperature °C
15.2	—		—		- 5.603	15.8	- 200
30.5	—				- 3.378	28.4	- 100
39.5	0.000	5.25	0.000	5.4	0.000	38.8	0
40.5	0.141	6.0	0.142	6.0	0.992	40.7	+ 25
41.4	0.647	7.5	0.645	7.3	4.277	46.8	+ 100
39.9	1.468	8.85	1.440	8.45	9.286	53.2	+ 200
41.5	2.400	9.75	2.323	9.1	14.860	58.1	+ 300
41.9	3.407	10.35	3.260	9.6	20.869	61.8	+ 400
42.6	4.471	10.9	4.234	9.9	—		+ 500
42.5	5.582	11.3	5.237	10.15	—		+ 600
41.9	6.741	11.8	6.274	10.55	—		+ 700
41.0	7.949	12.3	7.345	10.8	—		+ 800
39.9	9.203	12.7	8.448	11.2	—		+ 900
38.9	10.503	13.2	9.585	11.5	—		+1000
37.8	11.846	13.6	10.754	11.9			+1100
36.5	13.224	13.9	11.947	12.0			+1200
34.9	14.624	14.0	13.155	12.2			+1300
	16.035	14.1	14.368	12.2			+1400
	17.445	14.1	15.576	12.1			+1500
	18.842	13.9	16.771	11.8			+1600
	20.215	13.5	17.942	11.5			+1700
	—		—				+1800

∨—— R ——∨—— S ——∨—— T ——∨

°C/mV	°C	°C/mV	°C	°C/mV	°C	°C/mV	mV
	—		—		—		-10.000
43.48	—		—		-166.5	49.5	- 5.000
28.13	—		—		- 55.1	30.0	- 2.000
26.49	—		—		- 26.6	27.6	- 1.000
25.35	0.0	190.5	0.0	185.2	0.0	25.8	0.000
24.69	+ 145.0	122.7	+ 146.4	125.8	+ 25.2	24.6	+ 1.000
24.27	258.2	106.4	264.3	111.7	49.2	23.4	+ 2.000
24.42	548.2	90.1	576.6	99.0	115.3	20.9	+ 5.000
24.60	961.7	76.6	1035.8	86.2	213.3	18.6	+10.000
23.50	1684.1	73.5	—		385.9	16.3	+20.000
23.95	—		—		—		+30.000
25.48	—		—		—		+40.000
27.78	—		—		—		+50.000
							+60.000
							+70.000

3. The variation of tempcos from point to point provides a quick guide to linearity, both qualitative and quantitative. For example, the tempco of a Type J thermocouple changes by less than $0.7\mu V/$°C over any part of the range from 200°C to 500°C.

4. The tempcos are important data for many applications in which the *variations* of temperature are important, rather than the absolute magnitude. In addition, they can be used to obtain a direct indication as to how good the performance of the associated interface circuitry must be.

Temperature in °C has been used exclusively in order to avoid any possibility of misinterpretation. In addition, °C conforms to SI standards for temperature units; the conversion to Fahrenheit, where required, is universally known and understood.

Table 7-2 is a convenient cross-reference to °C, from millivolts of thermocouple output. Since all measurements provide millivolts, rather than degrees, this table will be a useful means of translation. Both temperature and incremental sensitivity (in degrees Celsius per millivolt) are given, to make interpolation easier. If plotted, degrees vs. millivolts, it provides a graphic view of linearity, since the departure of values (measured on the vertical axis) from a straight line represent both the nonlinearity and the amount of correction required.

Resistance Temperature Detectors (RTDs)

Chapter 8

The resistance change in RTDs caused by temperature (see Chapter 1) is sensed in two ways, either directly, as a change in voltage across a current-driven resistor, or by the output of a resistance bridge. The most-frequently used RTD is platinum; though it is expensive, its properties are stable and predictable. At the end of this chapter, abbreviated ready-reference tables provide the resistance and incremental temperature coefficients for a 100Ω (0°C) platinum resistor.

SIMPLE OP-AMP INTERFACE

Figure 8-1 shows a 100Ω RTD connected to perform temperature measurements in the range 0°C to 266°C, using simple, low-cost circuitry. The RTD is connected in the feedback path of an operational amplifier (A2); the 18mA excitation current is established by the net voltage across the input resistor. The 2.5V output of the AD580 reference IC (version a) is amplified to the 6.25V level (which permits a 6.2V zener diode to be used optionally, if sufficient current is provided). An excellent option is to replace the AD580 and AD741J by an AD584 adjustable multi-reference, set for 6.2V output at the external 2N2219 booster's emitter (b). Reference and output polarities are reversed.

The 1kΩ pot and the 50Ω variable resistor provide offset and span adjustments to set the output to 0V at 0°C and 1.8V at 266°C (first 1.8V span, then 0V offset). The scale is somewhat arbitrary, determined, in the case of the circuit shown here, by the voltage range required by the circuitry fed by this circuit.

The RTD's resistance varies from 100Ω to 200Ω over the temperature range. Best-straight-line linearity is to within about ±1.3°C.

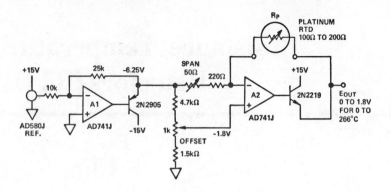

a. Low-cost fixed reference and op amp

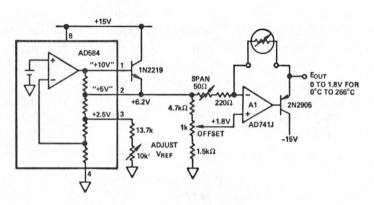

b. Boosted adjustable reference

Figure 8-1. Simple interface for RTD

USING A SIGNAL-CONDITIONER

In Figure 8-2, a Model 2B31 signal conditioner provides complete signal conditioning for temperature in the range –100°C to +600°C, using a YSI-Sostman four-wire, 100Ω platinum RTD (PT139AX).

The signal conditioner, functioning as a high-impedance current source, provides 1mA of excitation; the output voltage sensitivity is thus 1mV/Ω—about 350μV/°C. Because the device has four leads, one pair for excitation and the other for voltage pickoff, errors due to voltage drops are minimized.

The offset terminal of the signal conditioner allows the reference level to be shifted, and the span terminals provide for gain adjustment. In the example shown, the offset and span can be adjusted

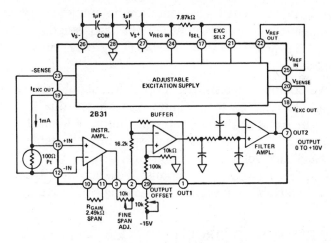

Figure 8-2. Platinum RTD in current-excited temperature measurement using 2B31 signal conditioner

for 0 to 10V output over the temperature range being measured. Measurement resolution and repeatability are ±0.1°C.

BRIDGE CONFIGURATION USING 3-WIRE RTD

As noted earlier, a bridge configuration is particularly useful for providing offset in interfacing to a platinum RTD, so that small, fractional sensor resistance changes can be detected stably and accurately.

In the configuration of Figure 8-3, an RTD is used as the active

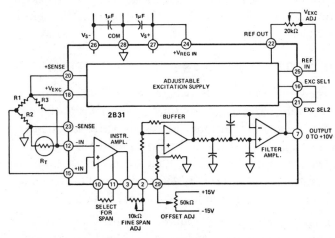

Figure 8-3. Platinum RTD measurement using resistance bridge

leg of a bridge. Lead compensation is employed to maintain high measurement accuracy when the lead lengths are so long that thermal gradients along the RTD leg may cause changes in line resistance.

The two completion resistors, R1 and R2, should have good ratio tracking (±5ppm/°C) to minimize bridge error due to drift. The resistor in series with the platinum sensor, R3, must have high absolute stability. The offset and span are adjustable independently, and the voltage excitation provided by the 2B31 can be adjusted for the best compromise between sensitivity (higher voltage) and stability (avoiding self-heating—lower voltage).

LINEARIZING RTD CIRCUITS

As tables 1 and 2—at the end of the chapter—show, platinum RTD's have a departure from linearity that is quite large compared to their resolution, stability, and repeatability. For example, over the range 0° to 558°C (100 to 300 ohms), the nonlinearity bow approaches 13°C (calibrated end points).

There are a number of ways to considerably improve the linearity of responses having a bow-shaped error curve. Since they have been discussed earlier, and are of universal applicability, the reader who is interested in this topic should consult Figures 5-9 and 5-11, and the associated text. They show how to linearize the output of bridges, using feedback around the bridge drive (in the former case) and analog multipliers (latter), *adjustably*, to permit correction for nonlinearity due to both the bridge and the sensor.

CURRENT TRANSMITTERS FOR RTD OUTPUTS

For process measurements, where temperature information must be communicated via a 4-to-20mA current loop, the voltage output of the signal conditioner is used to drive a voltage-to-current converter, such as the 2B20. In Figure 8-4, the output of a 2B31 provides the input to a 2B20, which can drive a loop with maximum resistance of 950Ω at +24V supply, within the specifications, and an absolute maximum load of 1.35kΩ, with a 32V supply.

In this application, ISA Standard 50.1, for Type 3, Class L and U, non-isolated current-loop transmitters, is met. If isolation is desired, the 2B22 V/I converter may be used.

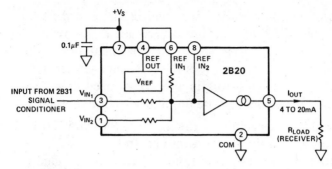

Figure 8-4. Current transmitter connection diagram

RTD-BASED PRECISION CONTROLLER

Figure 8-5 is a circuit that was developed to keep a small oven in a spacecraft at a temperature of 200°C ±0.1°C for *five years* under varying environmental conditions. A platinum RTD in a 3.5kΩ bridge circuit has a resistance of 3.5kΩ at 200°C. Any change in temperature will produce an unbalance voltage, which is amplified (A1), filtered (A2), and applied as a modulating signal to an oscillator (A3), which drives a resistance heater via a power FET.

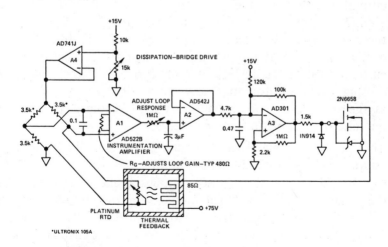

Figure 8-5. RTD-based precision temperature controller

The bridge drive is adjusted for the best compromise between output level (sensitivity) and dissipation (errors due to self-heating). Amplification is provided with low drift ($2\mu V/^{\circ}C$ max) by the AD522B, which also provides high common-mode rejection. The resistor in series with the platinum sensor must have high stability, because it is the resistance reference for the sensor; the other two resistors in the bridge need only be well-matched and tracking, and in close proximity to one another.

The $0.1\mu F$ capacitor at the input of the AD522 filters out noise spikes picked up at the front end from the hash generated by the switching circuitry. Additional filtering is provided by the RC at the input of A2; it also serves as the dominant pole in the feedback loop.

The output of A2 biases a pulse-width-frequency modulator. The more negative the output of A2, the greater the average power furnished to the 85Ω heater. When the output of A2 is positive, signifying that the temperature is high, the output of A3 is driven negative, turning the FET off and biasing A3 via the positive-feedback divider. If the output of A3 goes negative enough to bring the negative input of A3 below the threshold at the positive input, the output switches positive, applying power to the heater. The positive output also switches the input threshold positive and charges the $0.47\mu F$ capacitor, at a rate determined by the net current supplied to the capacitor via the $4.7k\Omega$, $100k\Omega$, and $120k\Omega$ resistors (the more negative the output of A2, the slower the rate). When the capacitor's voltage crosses the threshold, the output of A3 switches negative, changing the polarity of the current through the $100k\Omega$ resistor, turning off the power, and causing the capacitor to charge negatively. The cycle repeats; as the oven heats and the output of A3 becomes less negative, the ratio of time off to time on increases.

The modulated high-speed switching provides the benefits of continuously controlled temperature without perceptible temperature variations due to the discontinuous application of power; at the same time, the switched mode of power delivery provides efficient operation, essential for the application. The stability of temperature in the oven, with time and ambient change, is due to the stability of the bridge components and the amplifier.

MULTI-CHANNEL RTD THERMOMETER
The AD2037 is a 6-channel scanning* 3 1/2-digit integrating digital

*Its digital properties (display and scanning scheme) are the same as those of the AD2036, described in relation to Figure 7-10.

panel meter with a floating input system. In its simplest application, it can be used to monitor continuously millivolt-level voltages at six different points within a piece of equipment or a system. The basic full-scale range is 1.999V, but to make it easy to obtain offsets and other values of gain, it has a built-in operational amplifier, with a nominal gain of ten and all active terminals available. Thus, a full-scale range of 199.9mV is available, by simple jumpering.

In the application shown in Figure 8-6, the AD2037 is used to read temperature in the range 0° to 199.9°C, as measured by several 100Ω platinum RTD's, with repeatability and precision to within 0.1°C.

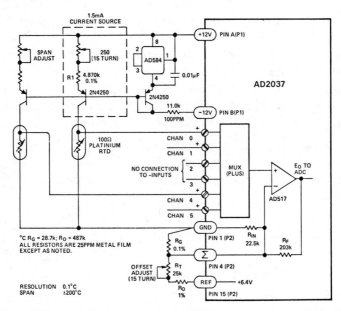

Figure 8—6. Application of the AD2037 with platinum RTD's

In this application, individual 1.5mA currents are applied to each device, and the voltages developed across them are multiplexed, applied to an amplifier, converted to digital, and displayed. The amplifier provides offset and gain so that the meter reads directly in °C.

The 1.5mA constant-current sources consist of 2N4250 transistors, 4.87kΩ resistors, and 250Ω rheostat-connected potentiometers to set the exact full-scale span. The reference for the excitation currents is provided by the AD584 multiple-reference IC, connected as a current source, in series with a 2N4250 and an 11kΩ resistor.

The resistance values shown are calculated for the proper relationship to obtain a full-scale reading of 199.9 to correspond to 199.9°C. The calculation process is described here to help you calculate values for different spans and resolutions.

Since the full-scale change of the RTD is from 100Ω to 175.84Ω from 0°C to 200°C, the span of resistance change for readings from 0° to 199.9°C is 75.75Ω. With an excitation current of 1.5mA, this corresponds to a voltage change of 113.625mV; therefore, the amplifier gain, for 1.999V full scale, must be 17.59.

The initial resistance of 100Ω at 0°C will produce an output of 100Ω × 1.5mA × 17.59 = 2.6385V, which must be offset by an added constant. The added constant is applied from the internal 6.4V reference, using an attenuation of –2.6385/6.4 = –0.41227. Since the nominal feedback resistance is 203kΩ, the input resistance must be 492.4kΩ. This value is achieved with a fixed resistance, $R_O = 487$kΩ, and a 25kΩ pot, R_T, to allow it to be trimmed.

To obtain the correct amount of gain for the 200°C span, a resistance, R_G, must be connected from the summing point to common, with resistance value to satisfy the relationship:

$$G = 17.59 = 1 + \frac{203}{22.5} + \frac{203}{492.4} + \frac{203}{R_G}$$

The nominal value of R_G is thus 28.37kΩ. The exact sensitivity of each channel, at a specific value of temperature, is set by the 250Ω pot in series with the emitter of each 2N4250.

Since the meter is bipolar, it will read negative values corresponding to temperatures less than 0°C. However, because of the nonlinearity of the platinum resistor, they will not be normalized to the 0 to –200°C range. At –200°C negative full-scale, the resistance is 18.53Ω; thus ΔR for that range is 81.47Ω, compared to 75.84Ω for +200°C full scale. Because the maximum reading occurs at a ΔR of 75.84Ω, the lowest temperature that can be read is nominally –186.7°C, which corresponds to 24.20Ω, for a reading of –199.9.

PLATINUM RESISTANCE TABLES

The following tables present, in compact form, data on the response of platinum RTD's to temperature, based on the resistance of a 100Ω device.

Table 1 lists, at 50°C intervals, the resistance (R_T), incremental resistance-temperature-coefficient, ($\Delta R/\Delta T |_T$), and the relative-resistance temperature-coefficient, $\left(\dfrac{\Delta R}{R_T} / \Delta T \right)$, from –200°C to +800°C.

PLATINUM RTD RESPONSE*

TABLE 1. RESISTANCE vs. TEMPERATURE

TABLE 2. TEMPERATURE vs. RESISTANCE (R_O = 100.0Ω)

TEMP °C	RESISTANCE (ohms — or millivolts per milliampere of excitation)	INCREMENTAL RESISTANCE TEMPCO $\Delta\Omega/\Delta°C$ (100Ω@0°C)	RELATIVE RESISTANCE TEMPCO $\Delta\Omega/\Omega/\Delta°C$ [%/°C]	RESISTANCE (ohms — or millivolts per milliampere of excitation)	TEMPERATURE °C
-200	18.53	0.421	2.27	20	-197.55
-150	39.65	0.416	1.05	40	-149.17
-100	60.20	0.406	0.67	60	-100.49
-50	80.25	0.396	0.49	80	- 50.63
0	100.00	0.391	0.39	100	0.00
+ 50	119.40	0.385	0.322	120	+ 51.56
+100	138.50	0.379	0.274	140	+103.98
+150	157.32	0.374	0.238	160	+157.19
+200	175.84	0.368	0.209	180	+211.33
+250	194.08	0.362	0.187	200	+266.39
+300	212.03	0.356	0.168	220	+322.46
+350	229.69	0.350	0.152	240	+379.57
+400	247.06	0.344	0.139	260	+437.79
+450	264.14	0.338	0.128	280	+497.21
+500	280.93	0.332	0.118	300	+557.85
+550	297.43	0.327	0.110	320	+619.84
+600	313.65	0.322	0.103	340	+683.26
+650	329.57	0.316	0.096	360	+748.19
+700	345.21	0.310	0.090		
+750	360.55	0.304	0.084		
+800	375.61	0.298	0.079		

*Summarized from The Omega 1979 Temperature Measurement Handbook, Omega Engineering, Inc., Stamford Connecticut 06907

In Table 2 are listed, at 20Ω intervals, the temperatures corresponding to given values of resistance. For 1mA excitation, the resistance column can be interpreted as millivolts of output. This table is especially useful in developing linearization circuitry, and in calibration.

Although the increments are relatively large, the relationship for platinum is sufficiently well-behaved that interpolation is easy, and surprisingly accurate, even with simple linear interpolation. The incremental resistance figures can be used for higher-order interpolation, as well as for sensitivity assessments.

Thermistor Interfacing

Thermistors are low-cost sensitive devices capable of operating over a moderate temperature range and available in a wide variety of standard resistance values (@ 25°C & negligible dissipation). They are discussed in Chapters 1 and 5. Many applications in measurement and control require reasonable degrees of accuracy and linearity; consequently, linearized composite thermistors, with guaranteed specifications, are preferred by many instrument designers to the lower-cost naked devices (which require more attention). Most of the applications discussed here involve linearized thermistors. Although one manufacturer's devices have received the lion's share of the mention, comparable devices are available from other thermistor manufacturers.

SIMPLE INTERFACE CIRCUITS

Linearized thermistors may be used in two ways: either as resistors, with resistance proportional to temperature, or as voltage dividers, with the ratio proportional to temperature. As expected, the former are two-terminal devices, the latter are three-terminal devices.

In Figure 9-1, a linearized thermistor is used as the feedback resistor of an op amp. A current, developed by an AD580 precision voltage reference and a series resistor, is transduced into voltage by the thermistor's resistance. The output voltage is summed passively with a constant offset and amplified. The net output voltage is proportional to temperature (10°C/V), over the range 0° to 100°C.

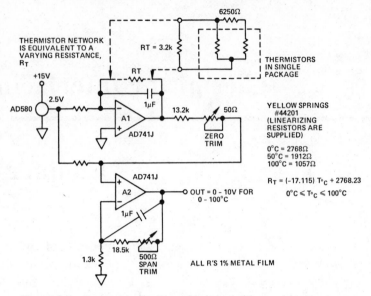

Figure 9-1. Instrumenting linearized thermistors with op amps—resistance mode

In Figure 9-2, which is functionally similar to 9-1, the same thermistor type is used potentiometrically. Both the sensor and the offset network are supplied by a 2.5V IC reference. The difference of the voltages (in essentially a linear bridge circuit) is read out by an instrumentation amplifier, which may be connected for the desired gain and output configuration (see Chapter Four).

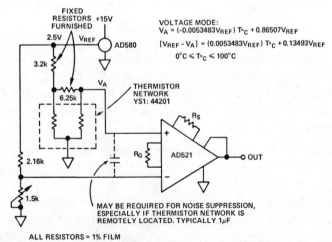

Figure 9-2. Instrumenting linearized thermistors—voltage mode

DIFFERENTIAL THERMOMETER

In Figure 9-3, two linear thermistor composites are used potentio-metrically to measure a temperature difference (for example, in a temperature control system, in gradient and thermal flow studies, in calorimetry, in chemical process monitoring, and a host of other operations). A 2B31 signal-conditioning module is used to provide excitation, precise differential-voltage measurement, amplification with adjustable gain, and (if needed) filtering.

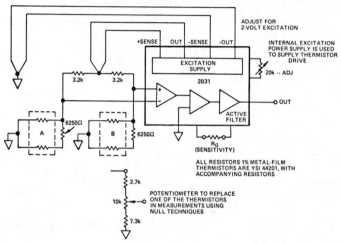

Figure 9-3. Differential thermometer using linearized thermistors

For precision temperature measurement using null techniques, one of the thermistors can be replaced by a ten-turn potentio-meter and associated scaling resistors (inset). At balance, the temperature can be determined from the potentiometer reading. The gain of the 2B31 determines the sensitivity of the null.

HIGH-RESOLUTION DIFFERENTIAL THERMOMETER

In the circuit of Figure 9-4, temperature differences can be mea-sured to tenths of a millidegree using a precision potentiometer—employing a Kelvin-Varley divider with 5-decade resolution; a floating chopper amplifier as a high-resolution null indicator; and an isolation amplifier to provide low-impedance readout at system level, as well as providing isolated power to float the chopper amplifier.

Two kinds of measurement can be performed: the differential temperature between thermistors A and B, and the temperature

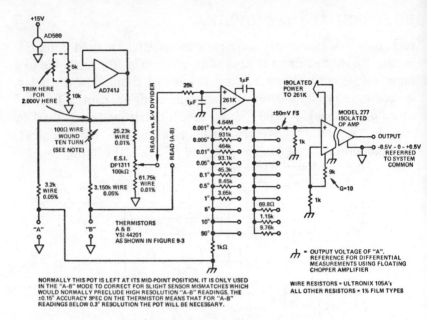

Figure 9-4. High-resolution differential thermometer

of thermistor A. The former is a precision measurement, similar to that shown in Figure 9-3, in which the 261K chopper amplifier reads the voltage across the bridge formed by the two thermistors. Because the 261K floats, it can perform differential voltage measurements with high common-mode rejection. The temperature at thermistor A is measured by forming a bridge with the precision divider, and adjusting the divider output for a null, again reading the output via the 261K. A number of gain steps are provided to permit the sensitivity of the measurement to be increased to the maximum needed. The 100Ω variable resistor is used to standardize the outputs of the two thermistors at the same temperature for measurements with better than 0.3° resolution.

Because the 261K floats across the bridge, it provides high common-mode rejection and 100nV/°C ambient drift performance, permitting extremely high gains to be used. On the most-sensitive gain setting, and when instrumented with all appropriate precautions, the circuit will allow stable resolution of 100 *micro-degree* (C) temperature changes at the sensor.

SMALL-DEVIATION THERMISTOR BRIDGE

Figure 9-5 suggests a means of measuring small (m°C) temperature changes. The active element in the bridge is a thermistor without linearization. A potentiometer, connected between the power-supply rails, provides an adjustable offset by injecting current into the fixed leg of the bridge, with offset sensitivity determined by a switched series resistor. Because the bridge excitation is provided by a pair of symmetrical or tracking supplies, the dc common-mode level is near-zero.

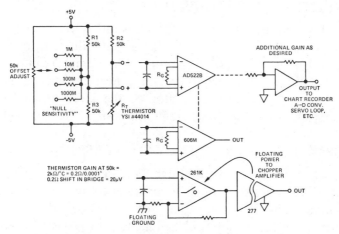

Figure 9-5. Alternative readouts for small-deviation thermistor bridge

At the initial temperature (near 60°C, in this case), the bridge is brought into balance by adjustment of the offset-adjust pot for zero output at an appropriate null sensitivity level. Then, when the temperature at the sensor has changed to the new value, the output voltage will be a linear function of the temperature change.*

For the YSI 44014 thermistor, which has a resistance of 50kΩ at 66°C, the temperature coefficient is approximately 2kΩ/°C (or 2Ω/m°C) at that point. The bridge can resolve temperature changes smaller than $100\mu°C$ when implemented and used with care (and a sufficiently sensitive amplifier). Typical amplifiers might include the AD522B IC instrumentation amplifier, the 606M low-drift instrumentation amplifier, and the combination suggested in Figure 9-4: the low-drift 261 chopper amplifier isolated by a 277—which provides gain, low-impedance readout, and isolated power. Because of the high resistance values, amplifier offset-current tempco is as important a consideration as offset voltage tempco.

*For doubled sensitivity, replace R1 by a matched thermistor.

Another application of this circuit might be to determine when the measured temperature has changed by a prescribed small amount. In that case, an offset would be preset, and the amplifier would be followed by a zero-biased comparator, to detect the crossing through zero difference between the preset temperature and the actual temperature.

CURRENT TRANSMITTERS

Thermistor measurements can be transmitted via 4 to 20mA (or other standard) current loops. The output voltage from a thermistor-bridge amplifier (see Figure 9-2, for example) can be applied to the input of a voltage-to-current converter, such as the 2B20 common-supply current generator or the 2B22 isolated current transmitter, described elsewhere in this book (see Figure 5-4).

THERMISTOR-TO-FREQUENCY CONVERSION

The output of a thermistor-bridge amplifier can be applied at the input of a voltage-to-frequency converter for high-noise-immunity transmission, with the possibility of optical or pulse-transformer isolation, and conversion to a digital code at the destination.

In Figure 9-6, the differential output of a bridge circuit employing a YSI 44018 linearized thermistor is amplified by an AD522 instrumentation amplifier and converted to frequency by a 452L 100kHz-full-scale V/f converter. The circuit has fast response, millidegree resolution, and a temperature range of 0° to 100°C, with accuracy to within 0.15°C. An analog output is available, normalized to 10°/V, and the frequency output may be optionally gated on and off by a level at precisely determined intervals, so

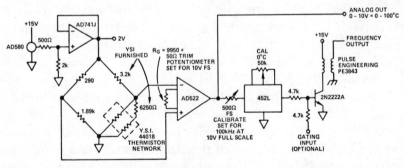

Figure 9-6. Thermistor-to-frequency converter

that a remote asynchronous counter may be used. The frequency output drives a pulse transformer, which provides galvanic isolation and off-ground operation.

LINEAR THERMISTOR THERMOMETER

A digital panel meter, such as the AD2026, can be used to read the output of a thermistor-bridge amplifier, simply and at low cost, in engineering units. Where more than one channel must be measured, the AD2037 six-channel scanning digital voltmeter, introduced in Figure 8-6, is a useful adjunct.

Figure 9-7 shows how the AD2037 would be connected to instrument a number of linearized thermistors (YSI 44201). Excitation for the thermistors is provided from the AD2037's reference voltage, attenuated to the 1.0000V level and stiffened by an AD741K op amp. The thermistors are read out potentiometrically and multiplexed by the meter's switching circuitry. Although the thermistors are excited in common and read with reference to the meter's input circuitry, they are isolated from the digital system circuits.

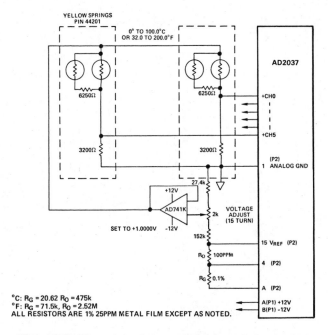

Figure 9-7. Thermistor instrumentation with the AD2037 scanning DPM

In this application, the temperatures being measured are in the range $0°$ to $100°C$ or $32.0°$ to $200.0°F$. The ranges are selected by the choice of resistors R_O (offset) and R_G (span), connected to the meter's internal precision op amp.

The digital properties of the AD2037, e.g., its conversion, display, and scanning schemes, are the same as those of the AD2036, described in relation to Figure 7-10.

Semiconductor
Temperature Transducers

Chapter 10

Low-cost semiconductor transducers are useful within the range, –55°C to +150°C. The types to be discussed here, as in Chapter 1, are *diodes, direct temperature-to-frequency converters* (the AD537), and *absolute-temperature-to-current converters* (the AD590).

T-TO-F CONVERSION USING DIODES

In Figure 10-1, the –2.2mV/°C change of voltage across the 1N4148 sensing diode modulates the frequency generated by a relaxation oscillator. The oscillator consists of an integrator and unijunction transistor, which periodically resets the charge across the 4300pF capacitor. The current through the capacitor, hence the rate at

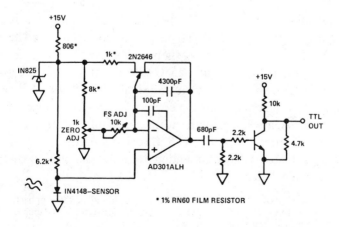

Figure 10-1. Diode temperature-to-frequency converter

which it charges (and the frequency), is determined by the difference between the voltage at the wiper of the zero-adjust pot and the voltage at the diode.

At the lowest temperature in the range, the zero-adjust pot is set as closely as possible for zero frequency. Then, as temperature increases, the voltage at the amplifier's + input decreases, and current flows through the 10kΩ pot and the integrating capacitor. The pot is set for full-scale frequency (about 1kHz) at full-scale temperature. For the range 0° to 100°F, accuracy is to within 1°F.

This approach is economical, in a sense; the concept and circuit are simple (but not the simplest), and the components are not expensive. However, each individual diode must be calibrated, and if a new diode sensor is installed, the circuit will require recalibration.

DIRECT T-TO-F CONVERSION WITH THE AD537

The AD537 (described in some detail in connection with the application shown in Figure 7-6) is an IC V/f converter, which requires a single capacitor to determine the frequency range. Either voltage or current can be used as an input. A unique device among VFC's, the AD537 has square-wave output (constant 50% duty cycle, irrespective of frequency), a 1.00V reference terminal (V_R), and a 1mV/K absolute-temperature-reference terminal (V_T).

The V_T output can be used as shown in Figure 10-2 to perform direct temperature-to-frequency conversion. It can also be used with other external connections in temperature-sensing or compensation.[1]

An absolute-temperature(kelvin)-to-frequency converter is easily implemented (a). The 1mV/K output is connected as the input to the buffer amplifier, which then scales the oscillator-drive current to a nominal 298μA at +25°C (298K). If a 1000pF capacitor is used, the corresponding frequency will be 2.98kHz. Adjustment of the 2kΩ trim resistor for the correct frequency at a well-defined temperature near +25°C will normally result in an accuracy to within ±2°C from −55°C to +125°C (using an AD537S*). An NPO ceramic capacitor is recommended to minimize nonlinearity due to capacitance drift.

---

[1] Many interesting applications of the AD537 can be found in the Application Note, "Applications of the AD537 IC Voltage-to-Frequency Converter", by Doug Grant. It is available at no charge from Analog Devices.

*Only the AD537S will perform to +125°C. For applications requiring maximum temperature of +70°C (+158°F), the J or K grades can be used with the same circuit values.

Other temperature scales are available by scaling and offsetting.
(b) shows connections for Celsius, and (c) shows the connections
for Fahrenheit.

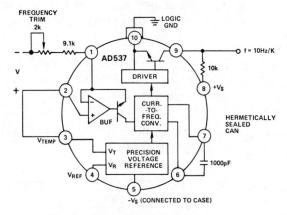

a. Absolute temperature to frequency

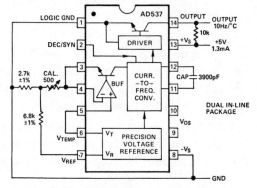

b. Celsius temperature to frequency

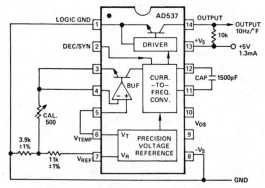

c. Fahrenheit temperature to frequency

Figure 10-2. Direct temperature-to-frequency conversion

For the Celsius scale, the lower end of the timing resistor is offset by +273.2mV, using the reference-voltage output. The component values take into account the required corrections for the loading of the 1.00V output (it is not zero-impedance). The frequency range in this application is 0 to 1250Hz, corresponding to 0° to +125°C.

The Fahrenheit scale requires an offset of +255.37mV, and increased gain (x9/5), produced by the component values shown. The output frequency range is 0 to 2570Hz for a temperature range of 0° to +257°F (–17.78° to +125°C), using the AD537S*. For variety, (a) shows connections for the hermetically sealed can version, (b) & (c) for the DIP version.

ABSOLUTE-TEMPERATURE-TO-CURRENT CONVERSION

The AD590, described in detail in Chapter 1, produces a current in microamperes that is numerically equal to the absolute temperature on the Kelvin scale (0°C = 273.2K), for temperatures from –55°C (218K or –67°F) to +150°C (423K or 302°F), independently of applied voltage over the specified +4V to +30V range. A two-terminal device, it is optionally available in a hermetically sealed TO-52 transistor package, a miniature flat-pack, a variety of probe hardware (AC2626), and in chip form (for user packaging or for measuring temperatures on circuit substrates).

The AD590 is available in several accuracy grades, as Table 1 indicates. The specifications of interest depend on whether the device is used uncalibrated or with calibration at a single value.

TABLE ONE. AD590 ACCURACY SPECIFICATIONS
(MAX ERROR)

Conditions		Max Error (±°C)			
Grade	I	J	K	L	M
Error at 25°C, as delivered	10.0	5.0	2.5	1.0	0.5
Errors over the –55°C to +150°C range:					
Without external calibration	20.0	10.0	5.5	3.0	1.7
With error nulled at 25°C only	5.8	3.0	2.0	1.6	1.0
Nonlinearity	3.0	1.5	0.8	0.4	0.3

For greater accuracy (in any grade), the device may be calibrated at two points. Since accuracy is a function of calibration error, linearity error, and operating range, there is a wide variety of

---

*Only the AD537S will perform to +125°C. For applications requiring maximum temperature of +70°C (+158°F), the J or K grades can be used with the same circuit values.

possible worst-case accuracy specifications, many considerably better than 0.05°C, based on the above. They are tabulated in the Appendix, "Accuracies of the AD590."

SIMPLEST READOUT, AN ANALOG METER

In Figure 10-3, the circuit consists simply of an AD590, a 0 to 500μA zero-adjustable analog meter, and a source of voltage (say, a 6V lantern battery). The meter will read temperature directly, to the accuracy determined by the meter and the device grade specification, with the error nulled at 25°C (298.2K). As long as sufficient voltage is available to supply voltage drops due to line resistance and maintain adequate excitation voltage at the device, the meter reading will be independent of the distance between the device and the readout.

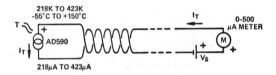

Figure 10-3. Simplest measurement system using the AD590 or the AC2626 probe

The AD590's breakdown voltage rating of ±200V, TO-52 case or stainless-steel probe to either lead, means that the AD590, or its probe version, the AC2626, may be placed in intimate contact to measure the temperature of conducting surfaces operating at common-mode voltages substantially higher than the AD590's supply.

Where the degree of resolution available is appropriate, the meter may be caused to read out in °C or °F, instead of K—without additional circuitry—by using an appropriately calibrated meter scale. If remote readout at several locations is desired, additional meters may be connected in a series loop.

VOLTAGE READOUT

In most applications, the AD590 will be interfaced to a system that requires a voltage input. In principle, all that is needed is a series resistance. A 1kΩ resistance (for example) in series with the AD590 will develop a 1mV/K voltage drop. This voltage may be applied at the input of an inverting or non-inverting op amp, a digital panel meter, or any other circuit capable of working with an accurately fixed resistive source. If readout of the same temperature is required at several remote locations, a current loop may be used with a (e.g.,) 1kΩ resistance (and either an instrumentation amplifier or a floating measurement system) at each location.

If the system that the AD590 feeds has offset and gain adjustments, a two-point calibration may be performed; calibrate the offset adjustment near the lowest temperature in the range, and set the gain (span) adjustment for null in the vicinity of the highest temperature to be measured. Single-point calibration for each AD590 used in a system is easily accomplished by trimming the load resistance at the specified temperature.

For readout in °C or °F, instead of kelvin, an appropriate amount of offset must be introduced, as Figure 10-4 shows. Also, for temperatures based on Fahrenheit, the nominal value of load resistance is 9/5 that required for temperatures based on Celsius.

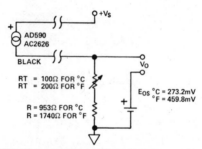

FOR MOST APPLICATIONS, A SINGLE POINT CALIBRATION IS SUFFICIENT. WITH THE PROBE AT A KNOWN TEMPERATURE, ADJUST R_T SO THAT V_O CORRESPONDS TO THE KNOWN TEMPERATURE.

Figure 10-4. Voltage readout using the AD590 or the AC2626 probe

The AD2040 is a digital panel meter designed specifically for low-cost single-channel digital-thermometer measurements employing the AD590 or the AC2626 probe. Available for either +5V system power or ac mains power (for isolated measurements), it provides all required offset and span calibrations for readout in K, °C, °F, or °R with ±1°C resolution.

Figure 10-5 is a block diagram of the AD2040, showing the AD2026 basic digital panel meter, the current-to-voltage span conversion resistors (R1, R2, R3), the offsetting resistor network (R4, R5, R6, R7), and the connections to the terminal strip. Attenuated voltage from the AD580 2.5-V reference provides the offsets for reading out the °F and °C scales. Jumpers are connected

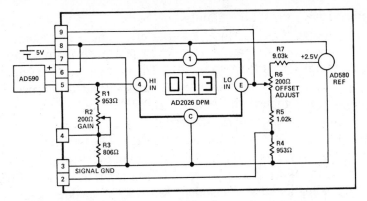

Figure 10-5. Connection diagram of the AD2040, showing internal connections and external connection to read Fahrenheit temperatures from -67°F to +302°F

by the user at the terminal strip, and the device is trimmed to read out in the appropriate units of temperature for display, according to this scheme:

Units	Connect These Terminals Together
Celsius	2, 3, 4
Fahrenheit	None
Kelvin	3, 4, 9
Rankine	3, 9

MEASURING DIFFERENTIAL TEMPERATURE

It is often necessary to measure, record, and control the temperature *difference* between two physically close—or widely separated—entities. A few examples of such temperature differences include the temperature rise above ambient inside a piece of equipment, temperature rise of a power transistor above ambient, the temperature difference between the inside and outside of an air-conditioning duct or plenum, and temperature gradients inside an oven or a refrigerator.

Depending on the application, the measurement of temperature differences may require resolutions as fine as a few millidegrees with full-scale ranges as great as several hundred degrees. For such applications, in the range from –55°C to +150°C, it is easy to implement ΔT measurements using the AD590 sensor or the AC2626 probe.

Figure 10-6 shows how simple the basic circuitry can be: two AD590's, two 9V transistor-radio batteries, and a microammeter. Currents I_1 and I_2 are generated independently in response to the temperatures of the two sensors. The currents flow in opposite directions through the meter, which reads their difference. For example, for temperatures of 35°C and 25°C, the current through the meter will be +10μA, corresponding to a difference of 10 kelvin, or +10°C. If I_1 is less than I_2, the reading will be negative. For perfectly matched IC's at the same temperature, the reading will be zero. If they aren't perfectly matched, zero can be trimmed by adjusting the mechanical zero of the analog meter. The meter can be protected by shunt diodes and an empirically determined value of series resistance (as shown in Figure 10-6) and/or by a shunting switch.

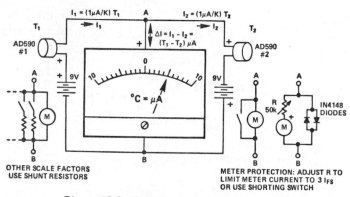

Figure 10-6. Differential temperature meter

If the meter has a 5″ scale, it can resolve changes of 0.1°C in a distance of 0.025″. For wider ranges of temperature change, a less-sensitive meter can be used, or range switching can be obtained by using resistive shunts across the sensitive meter. If greater sensitivity/resolution is desired, and one AD590 is always warmer than the other, a 10μA meter with 8″ scale length (e.g., Triplett 820 series) provides 0.025° resolution. Note that resolution is limited by the meter scale, not the errors of the AD590's.

For applications calling for voltage readout, the circuit of Figure 10-7a may be used. It demonstrates one way differential measurements can be made, using two AD590's and a single op amp. Resistors R1 and R2 allow offsets, due to either mismatches between the AD590's or to small temperature differences, to be trimmed out.

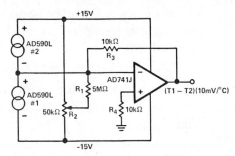

a. ΔT measurement with the AD590

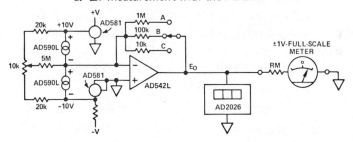

b. High-resolution ΔT measurement with AD590's or AC2626 probes.

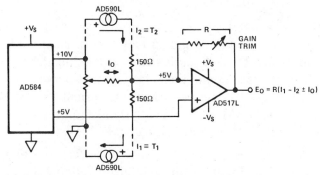

c. High-resolution ΔT measurement using the AD584 multi-reference to establish ±5V of excitation for the AD590's

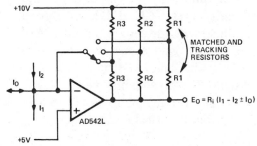

d. Gain-switching for (c)

Figure 10-7. Differential temperature measurement using the AD590

For even finer resolution, circuit b may be used. It adds range switching, the higher-performance parts required for 100X greater sensitivity, and readout devices.

Here's how it works: the AD542L TRI-FET* op amp, acting as a current-to-voltage converter, sums the opposing currents and converts them to voltage at a level determined by the feedback resistance. The $10k\Omega$ potentiometer is used to zero the output when the AD590's are at equal temperature.

In switch position A ($1M\Omega$ feedback resistance), the scale factor is $1V/°C$. The least-significant digit of the 999mV-full-scale AD2026 panel meter permits a resolution of one millidegree Celsius over the range from $-0.1°C$ to $+0.1°C$. Or, an analog meter can resolve 10 millidegrees over a $\pm1°C$ range. Resolution and scale factor for all switch positions:

	AD2026 DPM		Zero-Center Analog Meter	
Position	Resolution	Full Scale	Resolution	Full Scale
A	0.001°C	-0.099°C to +0.999°C	0.01°C	±1°C
B	0.01°C	-0.99°C to +9.99°C	0.1°C	±10°C
C	0.1°C	-9.9°C to +99.9°C	1.0°C	±100°C

The following precautions should be taken to insure good electrical performance at high sensitivities: use well-matched AD590's (L or M suffixes); excite them from stable dc voltage (AD581 precision references), and use a low-bias current op amp with low voltage drift (e.g., AD542L). The AD584 multireference permits simpler reference circuitry (c and d).

Figure 10-8a shows a simple circuit for differential temperature measurement employing a differential operational amplifier in a balanced-bridge single-excitation-source configuration. Output sensitivity is determined by the choice of R2 and R1; for example, $1k\Omega$ corresponds to $1mV/°C$, $1M\Omega$ corresponds to $1V/°C$.† R1 provides an adjustment for zero when both AD590's are at equal temperatures. If high gain accuracy is necessary, the differential measurement may be calibrated by a gain adjustment elsewhere in the signal path. The op amp may be an AD542 FET-input device for resolving small current changes or high accuracy at low cost, and an AD741 for non-critical measurements.

---

*Trimmed-Resistor Implanted-FET low-drift FET-input op amp
†Use shunt resistors, R_X, to reduce the common-mode voltage

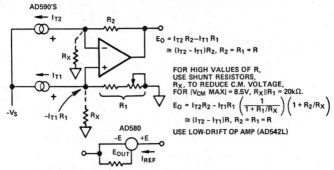

$$E_O = I_{T2} R_2 - I_{T1} R_1$$
$$\cong (I_{T2} - I_{T1})R_2, \ R_2 = R_1 = R$$

FOR HIGH VALUES OF R,
USE SHUNT RESISTORS,
R_X, TO REDUCE C.M. VOLTAGE,
FOR $|V_{CM} \ MAX| = 8.5V$, $R_X \| R_1 = 20k\Omega$.

$$E_O = I_{T2}R_2 - I_{T1}R_1 \left(\frac{1}{1 + R_1/R_X} \right) \left(1 + R_2/R_X \right)$$
$$\cong (I_{T2} - I_{T1})R, \ R_2 = R_1 = R$$

USE LOW-DRIFT OP AMP (AD542L)

REPLACE I_{T1} BY OPTIONAL TWO-TERMINAL CURRENT REFERENCE
TO ESTABLISH SET POINT $(E_O = I_{T2} R_2 - I_{REF} R_1)$

a. Single op amp

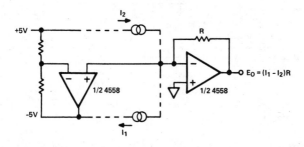

b. Dual op amp—CMV problem eliminated. A zero adjustment may be implemented in the same way as in 10-7c

Figure 10-8. Single excitation-source ΔT-to-voltage circuits

If one of the AD590's is replaced by a current source (e.g., an AD580 reference connected as a 1 to 11mA current source), then the output of the amplifier provides the difference between the measured temperature and a temperature set point, determined by the reference current and the value of its associated resistance.

AVERAGE AND MINIMUM TEMPERATURES

Since the AD590 is a current source, if a number of devices are connected in parallel, the total current flowing will be equal to the sum of the individual currents. Thus, a number of AD590's (n) may be used as a single composite sensor to measure the average of their temperatures. The load resistance will be R_L/n, where R_L is the resistance value used with a single AD590 (Figure 10-9a).

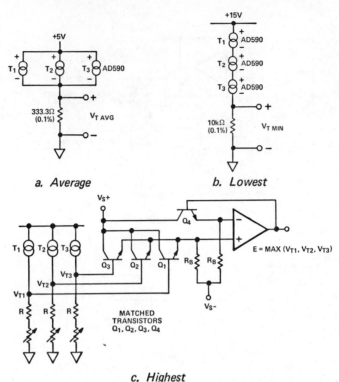

a. Average   *b. Lowest*

c. Highest

Figure 10-9. Combined measurements

If AD590's are connected in series, the current conducted will be limited to that of the device at the lowest temperature. Therefore, by the simple connection of Figure 10-9b, the lowest temperature in an ensemble can be measured. The *highest* temperature can't be measured quite as simply; the currents must be converted to voltage, and the voltages applied to an "auctioneer", or *upper selector*, an analog circuit that responds only to the highest voltage (c). The highest or lowest temperature at one point over a time interval can be computed by a peak- or a valley-follower (*Nonlinear Circuits Handbook*).

TEMPERATURE-CONTROL CIRCUITS

Figure 10-10 is an example of a variable on-off temperature-control circuit (thermostat) using an AD590. When the temperature at the AD590 is less than the set point, the output of the AD311 comparator swings to its upper limit, turning on the heating element. When the AD590's temperature is above the setpoint, the AD311 will turn the heater off.

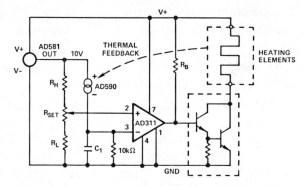

Figure 10-10. Simple temperature control circuit

R_H and R_L are selected to set the high and low extremes for the set point. Depending on the desired sensitivity and mode of adjustment, R_{SET} could be a simple pot, a calibrated multiturn pot, or a switched resistive divider.

The AD590 is insulated from power-supply variations by the 10V reference, for approximately constant dissipation, yet a reasonable voltage (about 6V) is maintained across it. C1 may be needed to filter out extraneous noise if the sensor is remotely located. R_B is determined by the β of the power transistor and the current requirements of the load.

Figure 10-11 is a schematic diagram of a continuously responding temperature-control servo that uses an AD590 as the sensor. The components to the right of A1 (similar to those in Figure 8-5) form a variable-pulsewidth modulator-oscillator to obtain smooth response despite the on-off nature of the heater excitation.

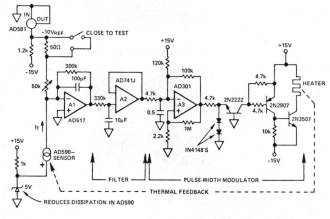

Figure 10-11. Temperature control system

The AD590's output current is compared—at the summing point of inverting amplifier, A1—with a reference current, generated by an AD581 voltage reference, connected as a high-performance –10V "zener diode". The AD517 sums the currents and amplifies the difference (error signal) to drive the servo.

The system for which this circuit was originally designed maintains the temperature inside a small cylindrical oven one inch in diameter, three inches long, and one-quarter inch thick. In order to obtain the most stable dynamic response in a configuration with tightly controlled mechanical parameters and simple dynamics, the AD590 is attached to the wall in close proximity to the heater winding via a film of silicone grease. The response of the error signal to a transient-response test (shorting out the 50Ω resistor in the reference circuit) is stable, as Figure 10-12a shows, b is the

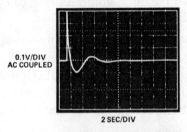

a. AD590 on heater coil with silicone grease

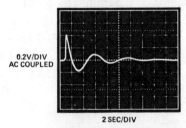

b. AD590 on heater coil

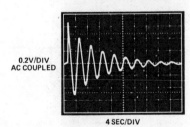

c. AD590 on wall near—but not on— heater coil

Figure 10-12. Response of temperature-control system

response to the same stimulus without silicone grease, and c is the response with the AD590 located a short distance away from the heater winding (lengthened time scale to permit more cycles of oscillation to be shown).

While it could be argued that this scheme does not actually control the temperature at the device inside the oven, thermal gradient measurements show that the temperature within the thick-walled, well-insulated oven is sufficiently uniform for low-dissipation loads to make the question moot. The lengthened thermal time constants that would be obtained by locating the AD590 within the oven would increase the problem of reliably obtaining and maintaining dynamic stability (and response speed) without substantially improving the accuracy with which temperature is maintained.

The setpoint of the control circuit in Figure 10-10 can be set digitally, using (for example) an 8-bit d/a converter. In the circuit of Figure 10-13, the desired temperature can be set to any value from 0°C (all inputs high) to +51°C (all inputs low) in 0.2°C steps. The comparator is fed back for a hysteresis band of 1°C, for reduced sensitivity to extraneous noise. The hysteresis can be avoided by removing the 5.1MΩ resistor.

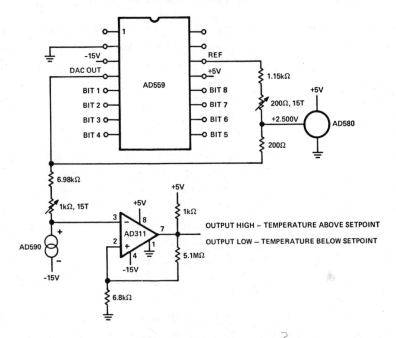

Figure 10-13. Digitally controlled setpoint

Figure 10-14 is an example of temperature control in which a system with a large amount of thermal lag is maintained at near-constant temperature with a high-capacity gas-fired heater by the use of carefully measured pulses of heat.

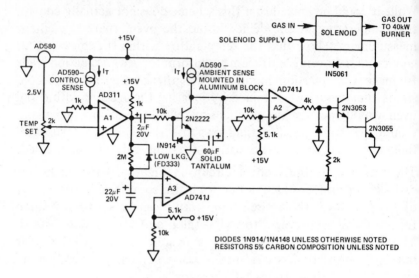

Figure 10-14. Temperature control system

If one assumes that the temperature has decreased sufficiently for the *control-sense* AD590's load resistor's voltage to drop below the reference voltage, the output of the AD311 comparator will swing positive, and the resulting pulse will turn on the transistor, discharging the 60μF capacitor. That negative swing will be inverted by the open-loop AD741, turning on the power transistors and the solenoid that turns on the burner.

The *ambient-sense* AD590, considered as a current source, will charge the 60μF timing capacitor positively at a rate of about 5V/s, faster at high ambients, slower at low ambients, until the threshold of A2 has been exceeded. Its output changes state and turns the burner off. A short time thereafter, the *control-sense* AD590's temperature-increase resulting from the burn will cause the output of A1 to go negative, clearing the ac-coupling capacitor and readying it for the next time the temperature has dropped below the set-point.

The function of A3 is to keep the burner on during startup, and to back up A2 during operation, if it should fail to switch. However, the long time-constant at the input of A3 keeps it out of action during the normally fast operation of the timer-A2 incremental-temperature-adjusting circuitry.

Circuits of this kind, carefully implemented, have been used to control a 5000-gallon vat to $100°C \pm 0.1°$ and to provide $0.001°C$ control resolution for an oven used to keep a quartz delay-line at constant temperature in a retrofit operation where heater and power-supply parameters could not be altered.

HIGH-LOW TEMPERATURE MONITORING

It is often desirable to have an indication when the temperature is higher or lower than a desired normal value by a given amount. The circuit of Figure 10-15 provides a pair of adjustable thresholds and LED indication of when they have been exceeded. This circuit can be used in conjunction with the digital readout provided by an AD2040 digital thermometer (Figure 10-5). The analog voltage developed across the (nominally) $1k\Omega$ resistance (for kelvin and Celsius), and appearing at terminal 5, is compared, in a CA3290E dual comparator, with the voltages set by the *hi limit* and *lo limit* pots, which derive their reference voltage from an AD580 2.5V reference.

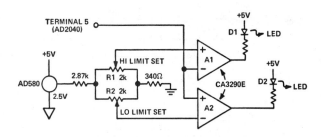

Figure 10-15. Limit detector and readout circuitry

When the voltage at terminal 5 goes higher than the *hi limit set* voltage, the output of A1 goes low and D1 is turned on. Similarly, when terminal 5 is at a lower voltage than the *lo limit set* voltage, the output of A2 goes low and turns on D2.

To set the high limit, using a meter that has already been calibrated, replace the AD590 with a variable resistor. Adjust the resistance until the reading of the meter is numerically equal to the desired high-temperature set point. Adjust R1 until D1 just turns on. Repeat the procedure to adjust R2 for the lower limit.

MULTIPLEXED APPLICATIONS

The high complicance voltage and high-impedance reverse blocking capability of the AD590 allow it to be powered directly from +5V CMOS logic, to permit easy multiplexing or switching, as well as pulsed measurements for minimum internal heat dissipation (or drain from the excitation/switching supply). In Figure 10-16, an AD590 connected to logic high (with a low gate input) will pass a signal current through the current-measuring circuitry, while the AD590's connected to logic zero (gates high) will pass insignificant current. The switch outputs used to drive the AD590's may be employed for other purposes, but the additional capacitance due to the AD590 should be taken into account, since it will cause some degradation of rise and fall times. The 1kΩ load resistance converts the absolute-temperature-proportional current to a voltage of 1mV/K.

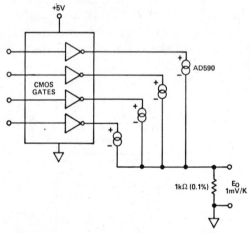

Figure 10-16. AD590 driven from CMOS logic

Figure 10-17 shows a method of multiplexing AD590's connected for high accuracy in the two-trim mode. As many as eight different temperatures can be multiplexed and measured to within ±0.5°C absolute accuracy over the temperature range –55°C to +125°C. Since the output of the op amp is limited with ±15V supply, to obtain accurate output over the full –55°C to +150°C temperature range, a +20V supply should be used for the op amp.

With a matrix approach, a large number of points can be monitored at a remote location using a small number of switches. Because the AD590's output is a current, switch resistance and line drop

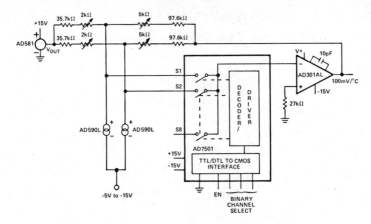

Figure 10-17. 8-channel multiplexer

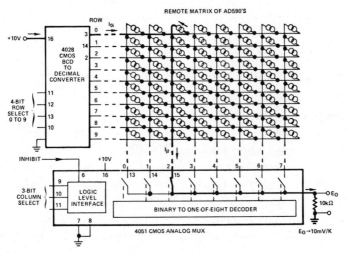

Figure 10-18. Matrix multiplexing scheme for temperature readout.
Heavy line shows path for I_{02}

are unimportant. The scheme shown in Figure 10-18 is a form of multiplexing uniquely applicable to the AD590. Here, the temperature of any one of 80 remote sensors can be read, independently of CMOS switch resistance, via only 18 wires, as addressed by a 7-bit word. The eighth bit can be an *inhibit* line that turns all sensors off for minimum dissipation while idling.

Figure 10-19 shows how an AD2040 low-cost digital thermometer may be used with a CMOS multiplexer to provide selective excitation and readout for eight AD590 sensors, according to the binary code applied to the address lines. The 10V Zener diode reduces the drop across the AD590's from 15V to 5V to reduce self-heating. Note that the sensors are shown schematically as current sources in this and other figures.

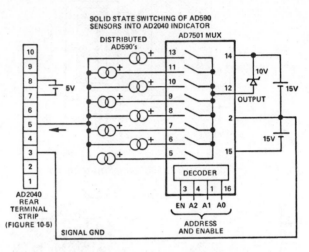

Figure 10-19. Selective excitation and readout among eight channels of temperature-sensing using a CMOS multiplexer

Multiplexing is not always electronic. Figure 10-20 shows a manual switching scheme for reading a variety of sources applied to an AD2040 digital thermometer. This figure also demonstrates the flexibility of powering the AD590. Here, it is shown powered (a) from the meter's supply, (b) from a remote lantern battery and a transistor battery. The meter itself is also shown being powered (in field applications) by a 6V lantern battery. Battery-excited AD590's should operate for a month on the 9V transistor battery

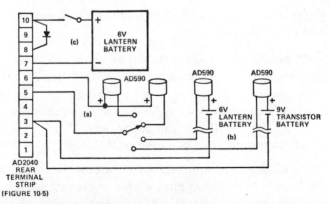

Figure 10-20. Typical modes of excitation for field applications, showing switch circuitry to select among sensors

or for one year on a 6V lantern battery; in intermittent operation, the meter should operate for 20 hours on the lantern battery.

The scanning digital thermometers, AD2036 and AD2037, have been mentioned earlier in association with thermocouple and RTD interfacing (see Figures 7-10, 8-6, and 9-7). A scanning thermometer dedicated to AD590 applications, the AD2038, is also available. In simplest terms, a single AD2038 provides the user with a tool to measure six temperatures in the range –55°C to +150°C (–67°F to +200°F) at six different remote points scattered over an area 5 kilometers (3 miles) square. Resolution is 0.1°C or 0.1°F. A twisted pair of insulated wire is all that is required to connect the measurement locations and the AD2038. Since the measurement is transmitted as a current, accuracy is independent of voltage drops, as long as the minimum voltage required for excitation appears across the sensor. The AD2038 provides the excitation, offset (for °C and °F), and readout.

Figure 10-21 shows the circuitry that provides the excitation for the reads the output of the AD590's. Each of the remote sensors must have at least 4V across it to function properly. The AD2038 can provide 7.5V of excitation, as shown for Channel 0. A remote battery (or a source of higher voltage) may also be used, as shown for Channel 5. Because there is no need to consider line- or contact-resistance at any point, the user may connect a variety of relays, switches, connectors, or even isolation resistors in series with the AD590 and AD2038 inputs, as long as sufficient excitation voltage is available.

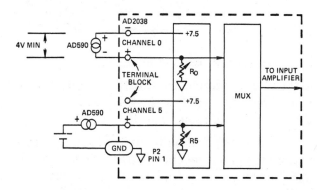

Figure 10-21. Applying the AD2038 six-channel scanning thermometer with AD590 two-terminal electronic temperature sensors

At the AD2038, the current from each AD590 is passed through an adjustable resistor (available to the user for trim) to converter to voltage for input to the AD2038's a/d converter and display.

ISOLATION

Temperature measurements are often performed in harsh environments where offground potentials, such as line voltage, may be accidentally impressed on the temperature sensor. In such cases, it is important to protect the sensor, the input circuitry, and (most important) the system electronics that the output communicates with. Two widely used forms of solution to this problem are isolation amplifiers (Chapter 4) and V/f conversion.

In Figure 10-22, showing one of perhaps several channels, an AD590 in a probe performs the current-output temperature measurement. A model 288 isolator and its associated 947 isolated supply provide isolated power for the AD590 and input/output isolation. A two-point calibration may be performed: calibrate $0°C$ by placing the probe in a zero-temperature bath and adjusting R_O for $E_O = 0V$; calibrate full scale by placing the probe in boiling water $(100°C)$ and adjusting R_S for $1.000V$ output.

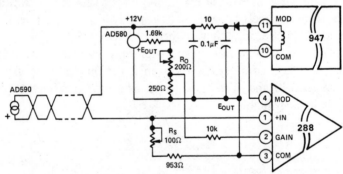

Figure 10-22. Isolated temperature measurements

Model 288 is a synchronized isolation amplifier, designed for multi-channel applications. For single-channel applications, Model 284 or 277 might be more suitable.

Figure 10-23a is an application where the simplicity of application and low power requirements of the AD590 combine with a monolithic V/f converter and an unorthodox power-transmission technique* to accomplish a difficult interfacing task.

*This circuit was reported to us by J. Williams, who advises us that a patent is pending on aspects of this system.

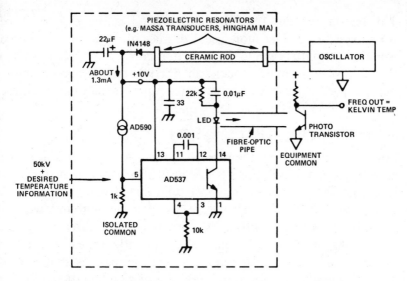

Figure 10-23a. 50kV-Isolated temperature measurement

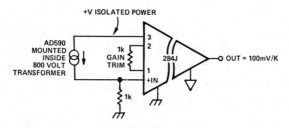

Figure 10-23b. Isolation amplifier in temperature measurement

A nuclear-physics experiment requires temperature monitoring of a 140°C surface (to 0.1°C resolution); the surface is floating electrically at *50kV!* The AD590 is mounted on the surface and its output serves as the input bias voltage for an AD537 IC V/f converter. Each time the open-collector output of the AD537 goes low, a quantity of charge is dumped through the 0.01μF capacitor into the LED, causing it to emit a short spike of light, which is transmitted via fibre optics. The electrical output is a frequency proportional to the temperature of the isolated surface.

Power for the AD590 and the AD537 is supplied by an oscillator, which drives one end of a ceramic rod acoustically, via a piezo-electric transducer. At the other end of the rod, another piezo-electric transducer converts the acoustic signal to an electrical signal, which is half-wave rectified to obtain dc power.

The reason for this unusual configuration is the high value of isolation voltage. For more-usual degrees of isolation, a standard isolation amplifier would be a better choice. For example, (b) shows how, in a similar circuit, an AD590 and an isolation amplifier are used to achieve contact temperature measurement in the presence of 800 volts of ac common-mode voltage.

4-20mA CURRENT TRANSMISSION

For some applications, the typical 300-microampere magnitude of current from the AD590 is either insufficient or is non-standard; and quite a few applications call for isolated current. For such applications, conversion to voltage, using an op amp, followed by a 2B20 voltage (0 to +10V) to current (4 to 20mA) converter, is a not-too-difficult solution.* Where isolation is called for, the 2B22 is useful, because it can accept voltage spans as small as 0 to 1V, and provide the appropriate degree of gain and offset, by the choice of external components.

For the designer who is measuring temperatures over a narrow range, the circuit of Figure 10-24 may be of interest. It converts the output of an AD590 to a 4-20mA range, in which 4mA corresponds to 17°C and 20mA corresponds to 33°C, for measurements of 25°C ambient ±8°C (i.e., the output sensitivity is 1mA/°C, a current amplification of 1000:1).

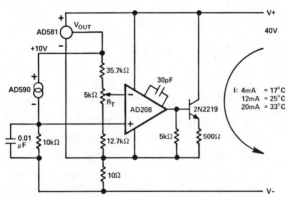

Figure 10-24. 4 to 20mA Current transmitter

SOUND-VELOCITY MONITOR

The thermometer and reference outputs of the AD537 V/f converter (Figures 7-6, 10-2, etc.) can simplify the measurement of

*Model 2B57 is a complete, two-wire, loop-powered, AD590-input, 4-to-20mA output transmitter.

many physical parameters. For example, the velocity of sound in air over a narrow range of temperature ($\sim 20°C$) can be computed from the formula:

$$c = (331.5 + 0.6T_C)$$
$$= (167.6 + 0.6T_K) \tag{10.1}$$

where

c is the velocity of sound in m/s
T_C is the temperature in degrees Celsius
T_K is the temperature in kelvin

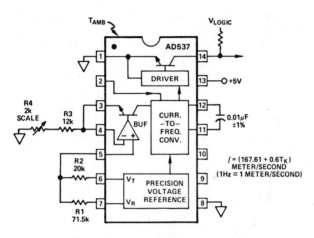

Figure 10-25. Sound-velocity monitor

In the circuit of Figure 10-25, R2 and R1 provide the appropriate weighting of the thermometer and reference outputs as given by:

$$(\text{const})\left[\frac{R_2}{R_1 + R_2} \cdot 1V + \frac{R_1}{R_1 + R_2} \cdot V_T\right] = 167.61 + 0.6T_K \tag{10.2}$$

From this, $R_2/R_1 = 167.61/600$. If R_2 is $20k\Omega$, then $R_1 = 71.5k\Omega$.

With these values, the voltage on pin 5 will be 452.8mV at 300K. This must be scaled to an output frequency of 347.6Hz, corresponding to the speed of sound at 300K. If a $0.01\mu F$ timing capacitor is chosen, the value of scaling resistance, R_3, is found to be $13k\Omega$.

Pressure-Transducer Interfacing

Chapter 11

This chapter considers applications of pressure transducers. Many types of pressure transducer are interfaced electrically in a manner quite similar to temperature transducers (for example, those using bridges—similar to RTD bridges—and potentiometers—similar to linearized thermistors configured as potentiometers).

Furthermore, pressure measurements are often transmitted as 4-to-20mA currents, or read out by scanning DPM's, or isolated (for safety or to handle large common-mode voltages), using the same facilities as temperature measurements. In such cases, the treatment is abbreviated, and the reader is referred to the earlier chapters, where more-complete discussions may usually be found. In any event, the reader will find familiarity with the temperature-transducer-application chapters—and, indeed, the introductory chapters—to be helpful for this chapter and those that follow.

STRAIN-GAGE-BASED TRANSDUCERS

Figure 11-1 shows an instrumentation-amplifier approach to signal-conditioning the output of a strain-gage bridge. Excitation for the bridge and for the amplifiers is furnished by the 2B35 supply,* set for 10V at the *sense* terminals, which monitor the voltage directly across the bridge. This compensates for voltage drops in the excitation line.

The bridge output is read by an AD522 IC differential instrumentation amplifier, with a gain of 1000, to achieve a 0 to 10V

*The 2B35 has a ±15V dual output, for amplifiers and electronic peripherals, and a programmable voltage or current output for transducer excitation. Input is at mains voltage and frequency.

output for zero to 100psi at the pressure transducer. Since the pressure is not expected to change rapidly in this application, a 10-second unit-lag filter is used for noise reduction, unloaded by a follower-connected AD542J FET-input op amp.

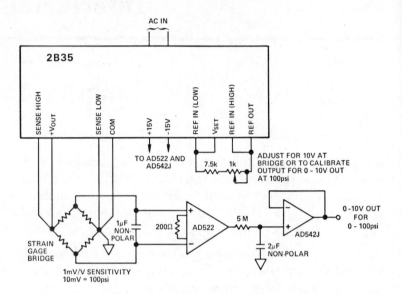

Figure 11-1. Pressure measurement interface with strain-gage bridge

The choice of AD522 grade depends on the ambient temperature range and the degree of resolution required. For example, the AD522A has a maximum offset drift of $6\mu V/°C$, referred to the input, at a gain of 1000. If the ambient temperature in the vicinity of the AD522 can vary by $±20°C$, the maximum drift would be about $±120\mu V$, a little more than 1% of full scale. The AD522B would have a corresponding maximum drift of $±40\mu V$, less than 0.5% of full scale. A 2B30 signal conditioner could have been used instead of the AD522 and the AD542, to provide a system solution.

Figure 11-2 shows the role of the 2B31 signal conditioner in interfacing a strain-gage pressure transducer (BLH Electronics, DHF Series). The 2B31 supplies regulated excitation (+10V) to the transducer and operates at a gain of 333.3 to achieve 0 to 10V output for 0 to 10,000psi at the pressure transducer.

The bridge-balance potentiometer is used to cancel out any offset which may be present, and the fine-span potentiometer adjustment is used to set the full-scale output accurately. A rapid check on

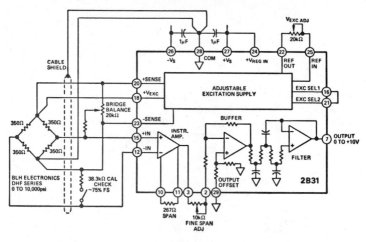

Figure 11-2. Pressure-transducer interface application using the 2B31 signal conditioner

system calibration can be obtained by depressing the *cal check* pushbutton switch, to shunt a calibration resistor across the bridge, which provides a reading of about 75% of full scale.

RHEOSTAT-OUTPUT PRESSURE TRANSDUCER

In transducers of this type, resistance is varied in proportion to pressure. If a constant current is caused to flow through the resistor, the voltage developed across it is proportional to pressure. The system-solution that can be implemented is to use the 2B31 to provide current excitation, amplification, and filtering, in the manner shown for RTD's in Figure 8-2.

Some simple circuitry, employing IC's to perform similar functions, is shown in Figure 11-3. In a, the $2.5k\Omega$ resistance element is furnished with current by an AD580 2.5V bandgap voltage reference connected as a current source ($I \cong \frac{2.5V}{R_L} + 1mA$).

Since $V = I \cdot R$, the voltage across the resistor measures its resistance, hence the pressure. The voltage is filtered and unloaded by an output follower. For the values shown, if $I = 2mA$, E_O will be 5.0V full scale, corresponding to 10psi/V.

In b, an AD581 10V reference, connected as a −10V two-terminal "zener diode", furnishes a precise −10V at the input of an inverting op amp. The $2.5k\Omega$ input resistor produces a feedback current of −4mA. The resistance element is connected in the feedback

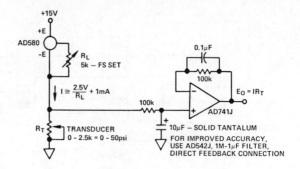

a. *Voltage reference as current source*

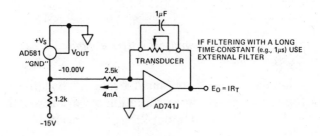

b. *Op amp as current source*

Figure 11-3. Simple circuits for instrumenting rheostat-output pressure transducers

path, so that the output voltage is proportional to its resistance. In this case, for 2.5KΩ full-scale resistance, full-scale output voltage will be 10V, corresponding to 5psi/V.

The resistance element discussed here reads zero ohms for zero psi. If there were an offset, offsetting could be used as discussed in Chapter Five. Some filtering of noise is provided by the capacitor across the transducer. If additional filtering were desired, a single lag (or a more-complex filter) could be used, as discussed in Chapter Three.

POTENTIOMETER-TO-FREQUENCY TRANSDUCER

A potentiometer-output pressure transducer can be read directly with a follower-connected op amp; it can be offset in a bridge configuration, as noted in earlier chapters. When used with the AD537 V/f converter, it will produce a pot-position-to-frequency conversion, in the form of a one-chip interface which can be powered by a single +5V supply.

The circuit of such an interface is shown in Figure 11-4a. The pot receives its excitation from the AD537's one-volt reference, and the fractional output, α, is applied to the + input. Full-scale frequency, as noted earlier, depends on R and C, and a constant, K, which is a function of the loading on V_{REF}. For example, a 10kΩ pot will require a K-factor of 0.825, as the graph in b shows. The frequency equation is

$$f = \frac{K \circ \alpha}{10RC} \tag{11.1}$$

where α is the fractional displacement of a potentiometer having a total resistance greater than 3kΩ.

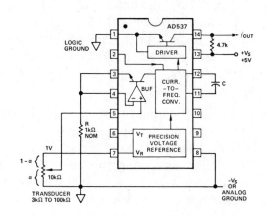

a. *Resistive-transducer-to-frequency interfacing*

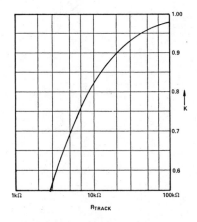

b. *K-Factor versus load on V_R*

Figure 11-4. Potentiometer-to-frequency transducer

As the equation indicates, with K = 0.825, R = 1kΩ, and C = 0.01μF, the nominal full-scale frequency (α = 1) is 8.25kHz. Polystyrene capacitors are preferred for tempco and dielectric absorption; polypropylene and Teflon are very good; polycarbonate or mica are acceptable; other types will degrade linearity. The capacitor should be wired as closely as possible to the AD537.

SCANNING PRESSURE METER

The AD2037 scanning digital voltmeter has already been discussed as a digital thermometer with RTDs and thermistors (Figures 8-6 and 9-7). Figure 11-5 shows how it is applied as a pressure meter. In this example, it furnishes excitation and 3 1/2-digit readout for a number of Data Instruments 0 to 100psi bridge-type pressure transducers. The meter's programmable gain and offset can be adjusted for a readout in any desired engineering units. The meter can scan automatically, manually or in an externally controlled random sequence.

The signal voltage appearing across the output leads of the transducer is a function of the applied pressure and the excitation voltage. The small residual offset voltage that is present at zero load can be nulled out using the zero-balance potentiometer. Transducer-span inaccuracies are calibrated via the span-adjust potentiometer. Since the front end of the AD2037 is differential,

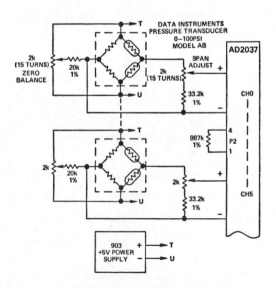

Figure 11-5. Multi-channel pressure meter

the difference measurement across the bridge will reject the 2.5V common-mode level, so long as analog ground is kept isolated from all parts of the pressure-transducer circuitry.

INTERFACING HIGH-LEVEL SEMICONDUCTOR TRANSDUCERS

The outputs of semiconductor transducers are high-level electrical signals that require little interfacing, except for offsetting and scaling. For example, the LX line of pressure transducers have a standardized output range of 2.5V to 12.5V corresponding to an input range from minimum pressure to maximum pressure.

Thus, any transducer in this family can be given a 0 to 10V range by subtracting 2.5V from the output. An AD580 and an analog subtractor is a low-cost approach (Figure 11-6a), requiring only four equal resistors. A summing inverter permits easy non-interacting offset and span adjustment, but inverts the input signal (b). Scale factors other than unity may be desirable if, for example, the 2.5V to 12.5V input is to be scaled to a 0 to 3V output to correspond to engineering units of 0 to 300psi (100psi/V).

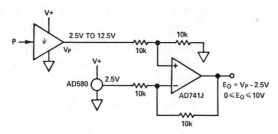

Figure 11-6a. Subtractor

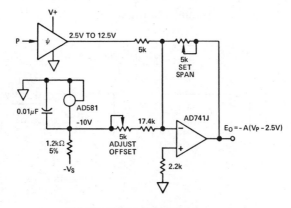

b. Inverting amplifier

The 12.5/2.5 ratio of max to min nicely coincides with the 20/4 max-to-min ratio of current transmitters. Therefore, a simple voltage-to-current transducer without offset will provide the appropriate current range. One scheme is shown in essence in c.

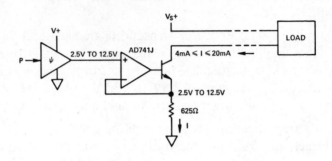

c. Rudimentary 4-20mA current transmitter

Figure 11-6. Interfacing semiconductor pressure transducers

ISOLATED PRESSURE TRANSMITTERS

The output of a pressure transducer can be amplified to a standard voltage level and then transduced to current for transmission in standard 4-to-20mA (or other such) current loops. The 2B20 provides a useful facility for non-isolated loops, and the 2B22 provides as much as 1500V of dc isolation, as Figure 11-7 shows.

Here, the Model 2B22 is used to provide complete input-output isolation and avoid signal errors due to ground-loop currents. The process pressure is monitored by a strain-gage pressure transducer interfaced by a 2B30 signal conditioner; and excitation for the bridge, the signal conditioner, and the V/I converter is

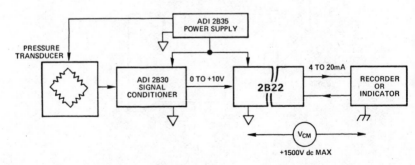

Figure 11-7. Isolated pressure transmitter

furnished by the 2B35 transducer power supply. The high-level voltage output of the 2B30 is converted to isolated 4-to-20mA current for transmission to a remote recorder or indicator.

PRESSURE-CONTROL SYSTEM

Figure 11-8 is a block diagram of a pressure-control system, showing how system-solution building blocks fit into a larger scheme. Here, the 3 to 15psi pressure in a tank is monitored by a strain-gage pressure transducer, interfaced with a 2B31 signal conditioner. The high-level voltage output of the 2B31 is converted to a 4-to-20mA current to provide a signal to the limit alarm and proportional control circuitry. A current-to-position converter controlling a motorized valve completes the pressure-control loop.

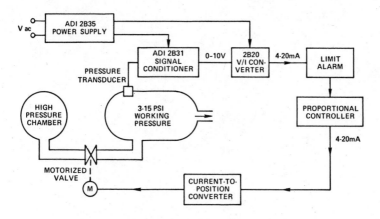

Figure 11-8. Proportional pressure control system

Force-Transducer Interfacing

Chapter 12

SPRING-DRIVEN RHEOSTAT

Some force-measuring transducers use a length of spring steel as the sensor. In the case of Figure 12-1, a spring is coupled to a rheostat. The resistance is proportional to the force applied to the spring, changing from 100Ω to 500Ω as the force increases from 0 to 20 pounds.

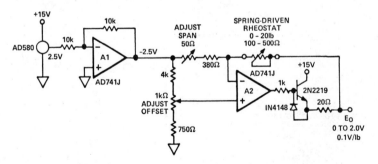

Figure 12-1. Spring-driven rheostat application

The circuit is very much like that used for the RTD in Figure 8-1. The variable resistance is connected in the feedback loop of an operational amplifier and driven by a constant current of 5mA. The 0V to 2V output range provides a scale factor of 10 pounds per volt. The reference signal is derived from an AD580 2.5V reference source, inverted in polarity in an AD741J op amp, and inverted once again in the output amplifier, to provide a positive output. The current-boost transistor provides for a load to be driven by the output amplifier.

The circuit is calibrated by adjusting the output span for 2V, for an input change of 20 pounds, then adjusting the offset for 0V at 0 pound-force.

STRAIN-GAGE LOAD-CELL INTERFACE

In this application (Figure 12-2), the sensor is a BLH Electronics T3P1 350Ω strain-gage-based load cell. The output of the bridge is amplified by an AD522 differential instrumentation amplifier. The bridge is excited by a current-boosted op amp, A1, the input to which is the circuit's 10V precision reference, an AD581. The entire circuit is powered by a Model 925 power supply, which is capable of supplying 350mA, enough for half-a-dozen similar transducers.

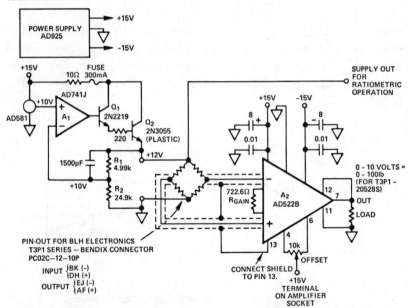

Figure 12-2. Load cell application with differential instrumentation amplifier

The op-amp circuit is connected for a closed-loop gain of 1.2 $\left(= 1 + \dfrac{R_1}{R_2} \right)$, so that the voltage at the output terminals is 12.0V.

The voltage at the bridge terminals is made available for referencing the bridge output to its actual drive voltage ratiometrically at a later stage (a/d converter, panel meter, processor, etc.)

In the booster circuit, the 10Ω resistor provides current limiting in the event of a short circuit. Since the follower is inside the feedback loop, variations of its gain do not affect the precision of the output voltage.

The sensitivity of the T3P1 load cell is 3mV per volt of excitation; with 12.0V applied, the full-scale output is 36mV. If the full-scale amplifier output is to be 10V, corresponding to 100 pounds (10 lb/V), the gain must be 10/0.036 = 277.8V/V. Using the AD522 gain equation,

$$G = 1 + \frac{200,000}{R_G} \qquad (12.1)$$

hence, the nominal value of R_G is 722.6 ohms. Since the gain equation is accurate to within 0.2% for the AD522B at 25°C, a nearby fixed value, a fixed-value-plus-fixed-trim, or fixed-value-plus-variable-trim may be used, depending on the gain accuracy required.

The AD522 has a guard-drive terminal (13) that follows the common-mode level. It may be used to drive short lengths of shielding on the input leads and the gain resistor. If the leads to be shielded are lengthy, the impedance at pin 13 should be stiffened by using it to drive a follower, which in turn drives the shield circuitry.

This circuit may be used to drive 120Ω bridges (e.g., BLH type V-1), which require lower drive potentials, for example 6.0V. For such applications, the AD581 output is attenuated to 6V, and the feedback from the bridge-drive voltage is unattenuated. The 2N3055 will have to dissipate about 0.5W, in this instance, vs. about 0.1W in the circuit of Figure 12-2.

STRAIN GAGE AND SIGNAL CONDITIONER

Figure 12-3 shows how a system-solution signal conditioner and transducer power-supply are used to provide excitation for and readout from a strain-gage bridge. In this case, the excitation for both the 2B30 signal conditioner and the bridge are furnished from mains power by a 2B35 transducer power supply. A single active gage (120Ω, G.F. = 2) is used in a bridge configuration with a dummy gage (for temperature compensation) and two precision 120Ω resistors.

The adjustable power supply is set for 3V of bridge excitation, to avoid errors due to self-heating of the gage and the bridge elements. The sense terminals ensure that the voltage *at the bridge* is precisely 3V.

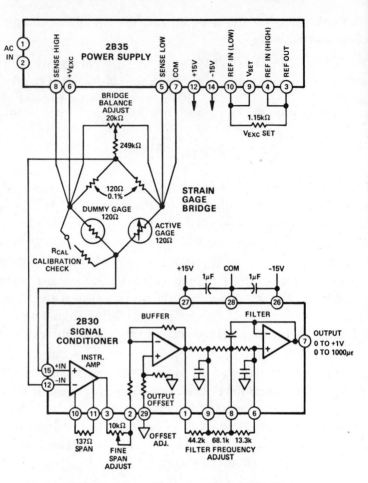

Figure 12-3. Strain measurement using a half-bridge and the 2B30 bridge signal-conditioner

The signal conditioner provides sufficient gain for an output of 1 volt, corresponding to 1000 microstrains. It also provides 3-pole Bessel filtering with a cutoff frequency of 100Hz.

The bridge-balance adjustment pot provides offset adjustment for the whole circuit, and fine span-adjustment is provided by a variable resistance connected between pins 2 and 3 of the module. A calibration-check resistance is shown across one leg of the bridge. It is of appropriate value to obtain a change in reading of about 75% of full scale when the switch is closed and opened.

Figure 12-4 shows a similar circuit employing the 2B31 signal conditioner, which contains its own adjustable bridge excitation and runs from system dc power. The circuit shows a 350Ω bridge, and a recommended shielding and grounding technique to help preserve the signal information.

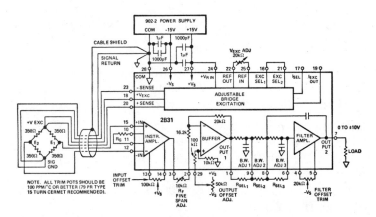

Figure 12-4. Typical strain-gage bridge transducer application using signal conditioner

Because these devices are direct coupled (i.e., not isolated), a ground-return path for bias current must be provided, generally through the bridge, as shown, unless the bridge is excited from a separate floating supply. If the inputs are floating, resistance up to $1M\Omega$ can be used to provide leaks from the supply ground to conditioner *common*.

For best performance, especially at high gains, the sensitive input and gain-setting terminals should be shielded from noise sources. To avoid ground loops, the signal return or cable shield should never be grounded at more than one point. The device should be decoupled from the power supplies, using $1\mu F$ tantalum and 1000pF ceramic capacitors as close to the amplifier as possible.

The 2B30 and 2B31 signal conditioners have been conservatively specified, using min-max and typical values, to allow the designer to develop accurate error budgets for predictable performance. An error budget for a typical transducer application, based on calculations for the circuit of Figure 12-4 using data-sheet specifications, is given here.

The circuit employs a 350Ω bridge, with 10V drive, and 1mV/V full-scale sensitivity (i.e., 10mV full-scale output). The assumptions used for the analysis are: Gain is 1000, ambient temperature range is $\pm10°C$, source unbalance is 100Ω, and common-mode noise at

60Hz is 0.25V (25 times as large as the full-scale signal) on the ground return.

Absolute gain and offset errors can be trimmed to zero. The remaining error sources and their effect on system accuracy (worst case) have been calculated. They are listed here as percentages of full-scale output (10V).

Error Source	Effect on Absolute Accuracy % of F.S.	Effect on Resolution % of F.S.
Gain Nonlinearity	±0.0025	±0.0025
Gain Drift	±0.025	
Voltage Offset Drift	±0.05	
Offset Current Drift	±0.004	
CMR	±0.00025	±0.00025
Noise (0.01 to 2Hz)	±0.01	±0.01
Total Amplifier Error	±0.092 max	±0.013 max
Excitation Drift	±0.15 (±0.03 typ)	
Total Output Error (Worst Case)	±0.24 max (±0.1 typ)	±0.013 max

The total worst-case effect on absolute accuracy over $\pm 10°C$ is less than ±0.25%, and this circuit is capable of 1/2 LSB resolution in a 12-bit low-level-input data-acquisition system. Since the errors are added directly, and the computation is based on conservative specifications, the typical overall accuracy error would be less than ±0.1% of full scale.

In a computer or microprocessor-based system, automatic recalibration can nullify gain and offset drifts; this leaves noise, non-linearity, and common-mode error as the only significant error sources. Transducer-excitation drift error is greatly reduced in many applications by ratiometric operation with the system's a/d converter.

HIGH-RESOLUTION LOAD-CELL PLATFORM INTERFACE

Figure 12-5 is the schematic of the high-resolution metabolic scale discussed in Chapter Six. The bridge is driven by a fully floating reference supply, employing a dc-to-dc converter. The ground-referenced 261K chopper-stabilized amplifier takes a highly stable true-differential measurement, with respect to the bridge output. Since the grounds of the two supplies are separated only by the bridge resistance, the 261K's bias current has a low-impedance ground return.

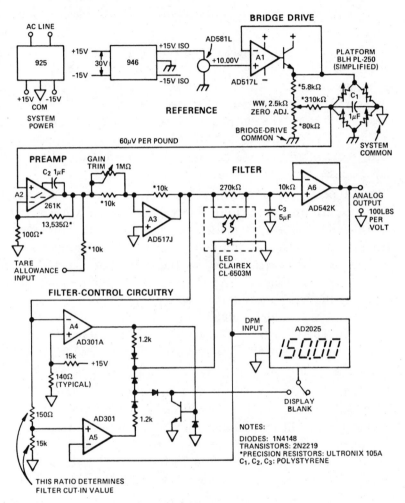

Figure 12-5. Precision platform balance circuit

The filter, in this interface, must eliminate very low frequencies (heartbeat, floor vibration), but must also respond quickly when someone gets on the scale. A simple low-pass filter with a 0.2Hz time constant will successfully eliminate noise artifacts, but it requires an unacceptable length of time to settle to five-place resolution. Higher-order linear filters will help, but they are more complex and are difficult to build and test at low frequencies. A simple solution is to employ an RC low-pass filter with a short time constant, and switch it to a long time constant after the scale has settled to within 99% of its final value. Since the time-varying filter used in this application is unusual, it will be described at some length.

With no weight on the scale, the output of A3 is at zero. The voltage at the + input of A5 is also at zero, and the voltage at its minus input is also very near zero because it is derived, via follower A6, from the same source as the + input voltage. Under these conditions, the output state of open-loop A5 is indeterminate. However, the output of A4, which is biased positive, is used to essentially ground any output from A5, via the transistor and diodes.

When a person or a weight of more than two pounds arrives on the platform, A4's output goes low (as A3's output goes high), and A5's output goes high, activating the light-emitting diode (LED). This permits rapid charging of the $5\mu F$ capacitor through the photoresistor. When the output of the AD542L is greater than the + input of A5, A5 goes to its negative limit, which shuts off the LED-photoresistor switch, after which the time constant of the filter is determined by the $270k\Omega$ resistor and the $5\mu F$ capacitor. This effectively filters body motion and heartbeat noise without the attendant long waiting time for a simple RC to settle.

When the subject steps off the scale, A4 goes high, the LED is activated, and the capacitor is drained rapidly towards zero. While the scale is empty, the output of A4 maintains the LED in the *on* condition, insuring rapid measurement response for lightweight objects or persons placed on the platform (i.e., those weighing less than 2 lbs). The LED-photocell switch provides drive-isolated, essentially errorless, switching for the application.

The requirements for the scale and the discussion of the low-level signal handling are discussed in detail in Chapter Six. To summarize, however, the scale is capable of reading 300.00 pounds full-scale (100 lb/V), with a resolution of 0.01 lb, and an absolute accuracy to within 0.05 lb. Developed at the Massachusetts Institute of Technology, it has been described in the technical literature.[1]

[1] Williams, J., "This 30ppm Scale Proves that Analog Designs Aren't Dead Yet," *EDN* Magazine, October 5, 1976, Vol. 21, No. 18

PIEZOELECTRIC TRANSDUCER SHAKER-TABLE CONTROL

In the application diagrammed in Figure 12-6, a servo loop drives a shaker table at constant amplitude at 400Hz. The acceleration/ force/amplitude of a shaker table is measured with a piezoelectric accelerometer. The resulting signal is amplified, and its rms value is computed and compared with a reference signal. The output of the low-frequency amplifier modulates the amplitude of a tuning-fork oscillator, and its output drives a transformer-isolated audio power amplifier, which excites the shaker table. If the table's amplitude tends to increase, the feedback signal tends to decrease it; if the amplitude tends to decrease, the result is increased drive, tending—in both cases—to maintain constant amplitude.

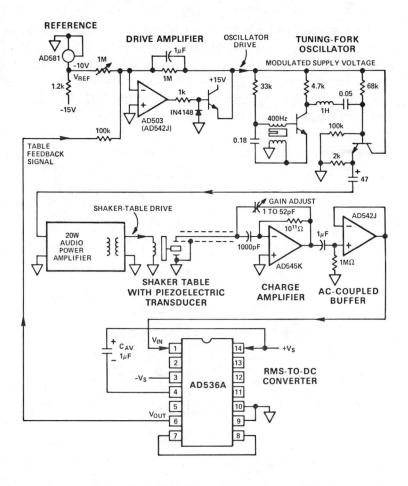

Figure 12-6. Constant-amplitude shaker-table drive

The output of the piezoelectric transducer can be thought of as a 400Hz voltage source in series with a small capacitor. Variations in capacitance due to physical deformation will cause charge to be dumped into the summing junction, and the amplifier will manipulate the output voltage to maintain the charge in the feedback capacitor equal to the charge dumped into the input, so that the voltage at the summing point tends to be at a null. The 1000pF capacitor and the large feedback resistor provide (1) ac coupling to avoid placing a dc bias voltage on the transducer and (2) a dc feedback leakage path for bias current to prevent the output from being driven into saturation.

The output voltage, as a function of Q(t) is

$$E_{OUT} = Q(t)/C_F \qquad (12.2)$$

where C_F is the feedback capacitance and Q(t) is the ac variation of charge developed by the piezoelectric crystal. The output of the charge amplifier is ac-coupled to the AD536 rms-to-dc converter, buffered by a follower-connected op amp. The output power of the driver op amp is boosted by the 2N2219 follower-connected transistor inside the local feedback loop.

STRAIN-GAGE TO FREQUENCY CONVERSION

The AD537 voltage-to-frequency converter, described in Chapter Seven and elsewhere, can be used for strain-gage-to-frequency conversion. Figure 12-7 shows an application in which the AD537's 1.0V reference source is used as the excitation for the bridge,

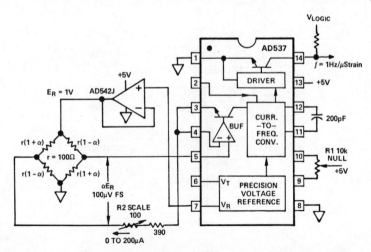

Figure 12-7. Typical connections for strain-gage-to-frequency converter

buffered by an AD542J FET-input op amp. In such applications, it is worth noting that, since the AD537's reference source is also the internal reference for the V/f converter, the conversion to frequency is ratiometric, hence quite stable with temperature and independent of the supply voltage.

The circuit of Figure 12-7 is calibrated to generate a scale of 1Hz per microstrain (100kHz at the assumed full-scale value of $\alpha = 0.1$). The assumption is that the strain is always of the same polarity (positive at terminal 5) and that the balanced set of 4 variable elements form a linear bridge. The timing resistor is returned to one side of the bridge, to both double the sensitivity and effect offset operation. Although this connection can introduce a nonlinear error term, for $R \geqslant 5r$ and $\alpha \leqslant 0.1$, the error is less than 0.1% absolute; if the system is calibrated at full-scale strain, this error is reduced to approximately +0.025% peak.

SCANNING STRAIN METER

The AD2037 scanning digital voltmeter, mentioned in previous chapters, is capable of providing excitation and readout for the broad range of transducers based on the ubiquitous strain gage. Applications include direct connection to a one-, two-, or four-gage bridge, and any of the transduction functions which use it, such as force, acceleration, level, weight, pressure, and flow. An example of its use in a bridge application is seen in Figure 11-5.

ISOLATORS AND TRANSMITTERS

As earlier chapters have noted, the outputs of interface circuits can be easily transformed to 4-to-20mA signals for transmission in industrial-process environments, with either common system wiring or isolation. Also, isolators and isolated op amps are available for two- or three-port voltage-to-voltage isolation. The AD2037 scanner, mentioned above, is powered by ac mains power and also has an isolated front end.

Flowmeter Interfacing

Chapter 13

The flowmeters discussed in this chapter are a small sampling of the many types in use, but they involve a variety of interfaces, some of which have been discussed in earlier chapters.

DIFFERENTIAL-PRESSURE FLOWMETERS

This class is based on the square-root relation between the flow through a resistance (such as an orifice or a restriction in the flow path) and the pressure drop across it. Differential pressure is measured, and the square-root of the resulting signal is computed. The square root can be computed as an analog signal, for immediate use; or, in systems involving processors, the differential-pressure signal can be converted to digital, the computation performed digitally, and the data used as needed for transmission, display, or further processing.

Figure 13-1 shows a simple means of computing the square root for unidirectional flow, using an electronic analog multiplier/divider, the AD534 or the AD535. The relationship is, simply

$$E_{OUT} = \sqrt{10(Z_2 - Z_1)}$$

Note that, since the Z input is differential, the differential pressure may be established by applying the outputs of two pressure-transducer interfaces to Z_2 and Z_1; if a single DP device is used, Z_1 is simply grounded (for positive input at Z_2). If the net input is negative, the roles of Z_2 and Z_1 should be interchanged. If it is desired to compute the negative of the output, reverse the external

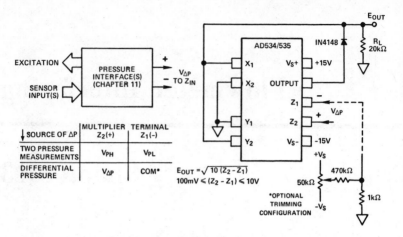

Figure 13-1. Square-root circuit for flow computation

diode connection and the X input polarity.

This circuit is normalized to 10V full scale. That is, for 10V in, representing ΔP_{max}, the output will be 10V. If the ratio of full scale to the lowest pressure is large, the optional trimming circuit shown should be applied to the normally grounded Z_2 input; alternatively, a pressure interface's *offset trim* can be used instead. The trim is adjusted for zero output as the differential input approaches zero.

FREQUENCY-OUTPUT FLOWMETERS

An important advantage of a flowmeter having a frequency output is that the signal can be transmitted with considerable noise immunity, despite little (if any) processing at the front end; also, the signal can be readily isolated and eventually transformed into either analog or digital information.

Figure 13-2 shows a circuit that has been used for instrumenting a turbine-type self-generating flowmeter. The 10mV output pulses are preamplified, using an op amp connected for gain of 1000, in an essentially ac-coupled circuit, to avoid amplifying offsets. The pulse frequency is converted to voltage in an f/V converter. The FVC is followed by a voltage-controlled low-pass filter, which further cleans up the waveform. The filter can also be adjusted to damp out the effects of periodic surges in flow, due to pumping, or to follow rapid slewing of the flow rate, depending on the control voltage applied. A follower-connected op amp buffers the output of the filter.

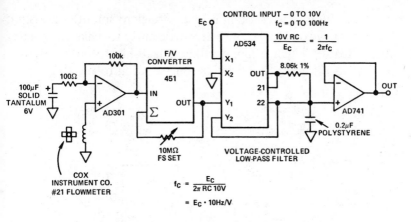

Figure 13-2. Flowmeter-to-analog interface

Figure 13-3, somewhat similar to Figure 13-2, shows an interface for a paddle-wheel-type flowmeter. A short pulse is produced each time a wheel arm passes the sensing point. The AD311 comparator senses the small pulses and provides logic pulses, which may be used to generate analog or digital information. For small signals, some gain (as shown) may be desirable, to allow the comparator to discriminate more easily.

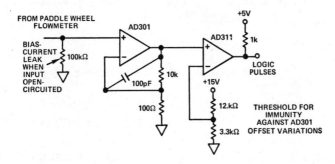

Figure 13-3. Interface for paddle-wheel-type flowmeter

ANEMOMETERS

The anemometer (*wind-meter*) is a common form of flow meter. In its most commonly seen form, the cup (or propeller) type, the output is a frequency, produced by actuating a reed switch magnetically for each revolution. This is a reliable scheme, with potentially long equipment life. However, if the readout is to be analog, the slow rate of rotation at low wind speeds poses the dilemma that if the filter time-constant is long enough to smooth the output of a frequency-to-voltage converter sufficiently to make the reading appear steady, it will respond rather slugglishly to puffs or gusts. The circuit of Figure 13-4 provides one solution to this dilemma.

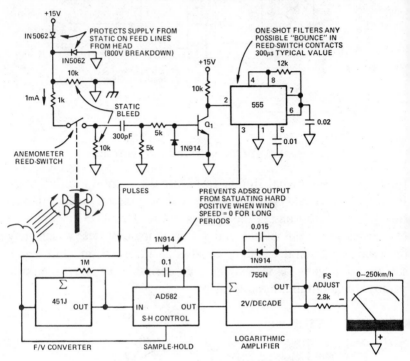

Figure 13-4. Anemometer circuit

In the system shown, the reed-switch closing drives Q1 into conduction, which triggers the 555 one-shot. The 300µs output pulse from the 555 is the input signal to the Model 451 f → V converter, with a 1MΩ feedback resistor for full-scale output at 100Hz. At high wind speed, the percentage of ripple in the output of the f/V converter is small, yet response is fast. At low wind speeds, the readings will have large, slow, unstable-appearing variations, unless heavy low-pass filtering is used.

The solution, in this system, is to use an AD582 sample-hold, which acts as a synchronous filter to sample the output of the 451J in synchronism with the input, then hold it until the next pulse. The output of the sample-hold is at a properly proportioned dc level, regardless of how low the repetition rate of the input signal is.

Better definition at low speeds is obtained by displaying wind speed logarithmically, using the 755N log/antilog amplifier and a meter with a log calibration.

The *hot-wire* anemometer measures speed of a medium by the cooling effect of the flow upon an electrically heated platinum filament (i.e., an RTD in the self-heating region). Figure 13-5 shows a typical voltage-current characteristic at two values of air speed.

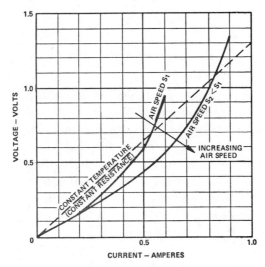

Figure 13-5. Volt-ampere characteristics of 7/16"L × 0.002"D straight filament of pure platinum in the presence of moving air, for two values of air speed.

When operated at constant current, the device is quite sensitive, but could self-destruct at very low airspeeds; when operated at constant voltage, the device is stable but less sensitive. In either case, a change in temperature is associated with the change in airspeed, hence a delay in measurement, which could be unstabilizing if the anemometer were part of a control loop.

Therefore, it is useful to operate the device at constant resistance (i.e., constant temperature), by adjusting the drive voltage (Figure 13-6)[1]. From the curves, one can see that, if the resistance were maintained at 1.3Ω, as indicated by the dashed line, one could obtain a current change of 0.3A and a voltage change of 0.4V, for the speed change shown, with no danger of overheating in

[1] Miyara, J., "Measuring Air Flow Using a Self-Balancing Bridge," *Analog Dialogue*, Volume 5, Number 1, 1971

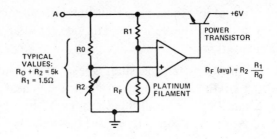

Figure 13-6. Basic circuit of the temperature-controlled bridge

normal operation. Response would be speedy, since temperature changes are transient and small.

In the circuit of Figure 13-6, the op amp continuously adjusts the flow of current to maintain the two inputs equal. This can be accomplished only by keeping the voltage across the filament equal to that across R2, and the filament current equal to the current through R1. However, since the current through R1 is proportional to the current through R0 (which has the same voltage drop as R1) and the current through R0 is determined by the voltage drop across R2, it can be seen that the resistance of the filament, R_F, must be equal to that of R2, multiplied by the ratio R_1/R_0.

Suppose that, starting from a given equilibrium point, the air flow increases. This will take heat away from R_F, causing its voltage to tend to drop. The amplifier's output voltage increases, which increases the current through the power transistor, and thus makes more power available for the filament to dissipate to maintain its temperature (and hence its resistance) constant.

The output voltage is measured at terminal *A,* which provides an amplified version of the filament voltage, at an impedance low enough to operate even the crudest of meter movements. The zero-speed voltage is biased off by an auxiliary constant voltage, and the readings can be displayed with a moving-coil meter. The scale is a nonlinear (calibratable) function of airspeed, actually expanding toward the lowest values. Thus, low air speeds can be read with high sensitivity; in fact, the device can virtually detect a whisper several feet away.

For practical realization, the following points should be considered:

1. A small keep-alive current must be introduced to insure that the output goes positive on turnon.

2. The power transistor must have ample current-handling capacity; the filament requires substantial fractions of 1 ampere.

3. Stability is dependent on the physical layout, especially if the filament is at the end of a twisted pair. The circuit should be checked with an oscilloscope, and appropriate measures taken to insure stability (capacitance from base to collector of the power transistor, small resistance in series with the base, compensating inductance in series with R1, etc.)

4. R0 and R2 form a trim potentiometer to set the operating temperature (e.g., resistance) of the filament. If R2 is a variable resistance, start with $R_2 = 0$, and increase it until the filament just starts to glow, then back down a little. This will give optimal sensitivity, as well as independence of ambient temperature.

Applications: The device has been used commercially to trip out equipment when the air speed in a forced-draft duct falls below a preset value, but the approach is highly suggestive of other applications in instrumentation.

HINGED-VANE FLOWMETER

The hinged vane (Figure 1-15b) can be linked to a potentiometer; the fractional rotation is a function of flow. In Figure 13-7, the pot is energized by an AD580 voltage reference, and the output is shown optionally feeding an AD2026 3-digit panel meter, a simple follower (and subsequent system instrumentation), and an instrumentation amplifier (where grounds are noisy).

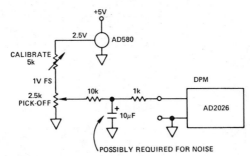

Figure 13-7. Interfacing a hinged-vane flowmeter

The calibration rheostat is adjusted for 1V full-scale across the potentiometer, or for a reading of 999 on the digital panel meter at (99.9% of) full-scale flow.

THERMAL FLOWMETER

If a heat source is thermally centered between two sensors, the difference between the temperatures they measure is a function of the flowrate (Figure 13-8a). Figure 13-8b is a simplified diagram of a patented flowmeter built upon this principle. The temperature sensors are AD590 semiconductor devices, and the difference between the voltages they develop across 1kΩ resistors is read out by an AD521 instrumentation amplifier. Heat is supplied by a controlled heating element (in this case, a vitreous resistor).

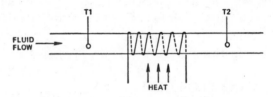

Figure 13-8a. Flowmeter principle

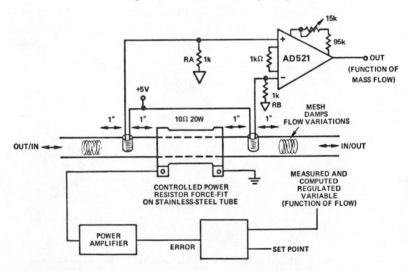

Figure 13-8b. Flowmeter circuit

TRANSMISSION, ISOLATION, AND READOUT

As earlier chapters have noted, the outputs of interface circuits can be easily translated to 4-to-20mA signals for analog transmission in industrial-process environments, with either common system wiring or isolation. Also, isolators and isolated op amps are available for two- or three-port voltage-to-voltage isolation. Frequency-output devices are easily isolated, optically or electromagnetically. Where a number of channels are to be read, the AD2037 6-channel scanning digital voltmeter provides a sensitive front end, 3 1/2-digit readout, and flexible digital control options.

Interfacing Level Transducers

Chapter 14

Level is usually measured to determine the *contents* of a container because of the linear relationship between level and volume (when the sides of the container are vertical). Contents can also be measured as mass (or weight), by weighing the container; and they can be determined by measuring pressure at the bottom of the container, if the density of the contents and the geometry of the container are known. Interfaces to pressure and force measurement are discussed in Chapters 11 and 12.

Level measurement can use extremely simple or very sophisticated techniques. For the latter, the electronic interface is usually part of the measurement package, and is not relevant for discussion here. A few examples of some simple, and quite common, techniques—and interfaces—are mentioned here.

FLOAT AND POTENTIOMETER

A float, linked to a potentiometer or a rheostat, is perhaps the most common form of level measurement. In such applications as vehicle fuel-tank level measurement, the level is read directly on a meter. If communication to a measurement system is desired, the simplest form of handling is via a follower-connected op amp.

In Figure 14-1, the potentiometer is excited by an AD581J 10V reference, and it is buffered by an AD741J op amp. The relationship between output and level can be expressed as 10% (of full height) per volt.

A plain (or fancy) RC filter may be interposed between the pot and the output of the follower to average out the effects of noise

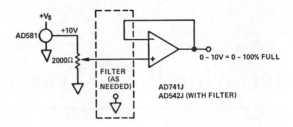

Figure 14-1. Level measurement using float and pot

(ripples, waves, and froth at the surface of the fluid). See Figures 3-7 and 3-8 and associated text for a discussion of some techniques. Generally, a FET-input amplifier, such as the AD542J, would be used if a filter is required.

An instrumentation amplifier, such as the AD521, or perhaps even an isolator, would be used if ground problems are expected (or suspected). For 4-to-20mA current transmission, the pot can be connected directly to the input of the isolated 2B22 V/I translator without loss of linearity; however, in the case of the common-based 2B20, a buffer should be used unless the nonlinearity caused by loading of the 10kΩ input impedance of the 2B20 is tolerable.

If, instead of (or in addition to) a direct analog reading, a pair of decisions must be made to turn a fill circuit on at one level and turn it off at another, the output of the pot (or the buffer) may be applied to one input of each of two comparators. The level-set signals (also derived from the AD581) are applied at the other inputs. The sense of the comparator connections (+ or –) is determined by whether the decision is *fill* or *stop*.

OPTICAL SENSING

A simple optical sensing system consists of a light source, a photo-detector, and appropriate circuitry. Figure 14-2 shows these elements. When the liquid level is low, the diode conducts, and the output of the comparator is high; when the liquid level interrupts the optical path, the output of the comparator goes low. If liquid motion causes excessive electrical noise, a filter can be inserted between the diode and the comparator input, as shown. A small amount of positive feedback from the output of the comparator could also be used for hysteresis to avoid frequent reversals due to low-frequency noise.

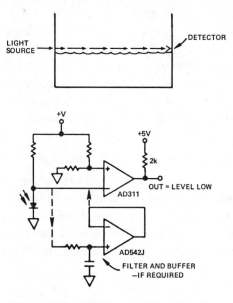

Figure 14-2. Simple optical level sensing

THERMAL SENSING

The self-heating characteristics of thermistors can be used to detect liquid level. In Figure 14-3, the 2kΩ (@ 25°C) thermistor has low resistance when it is hot (i.e., self-heated in air), and the output of an AD311 comparator goes high. When the thermistor is in contact with liquid, its temperature drops, its resistance rises, and the output of the comparator goes low.

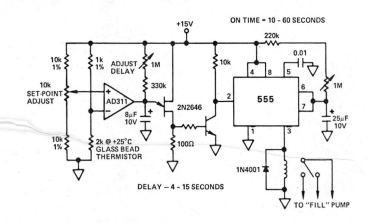

Figure 14-3. Thermistor-sensed level-control system

In a typical fill-control application, such as the one shown in Figure 14-3, when the liquid level drops away from the thermistor, the open-collector output transistor shuts off, and the current through the adjustable pullup resistor charges the $8\mu F$ capacitor up. When the 2N2646 unijunction transistor fires, it operates the 555 timer, which cause the *fill* relay to close. The 2N2646 continues to be turned on until the liquid rises to cool the thermistor.

As the liquid is drawn off and the level drops, this circuit will thus act to restore the liquid level to the set point.

The delay time in triggering the 2N2646, caused by the charging of the capacitor, is adjusted so that turbulence and sudsing action at the surface of the liquid do not cause the circuit to fire prematurely. The timer *on-time* is adjusted to provide enough hysteresis so that the circuit does not oscillate at the fill point.

Application Miscellany

Chapter 15

In this chapter, we discuss some circuit techniques and hardware that go beyond the basics covered in Part One, are useful for more than just one type of sensor, and embody some interesting ideas.

4-TO-20mA TRANSMISSION REVISITED

Circumstances can arise in which it is useful to tap into an existing current loop and transmit up-to-1500-volt-isolated 4-to-20mA information via a second loop. Figure 15-1 is a circuit for interfacing a process loop (having an arbitrary 5:1 current range, for example 1 to 5mA, 4 to 20mA, 10 to 50mA) with an isolated 4-20mA loop, using a 2B22 isolated V/I converter. Since the current range is proportional to that of the output, it is necessary only to transform the input current (I_{IN}) to voltage with a resistance, R_C, and to scale the input voltage range (V_{INFS}) to the out-

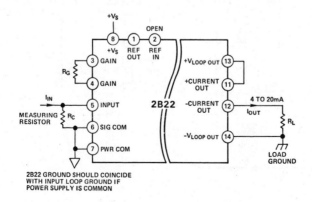

Figure 15-1. Process signal current isolator

put full-scale range, via resistance R_G. The formulas given for R_C and R_G are:

$$R_C = V_{INFS}/I_{INFS} \tag{15.1}$$

and

$$R_G = \frac{6.31 V_{INFS}}{12.62 - V_{INFS}} k\Omega \tag{15.2}$$

The value of R_G determined in (15.2) has a tolerance of ±5%. Therefore, R_G should consist of a smaller-than-nominal value in series with a trim resistance. V_{INFS} can be any value in the range 1V to 10V, as permitted by the primary loop.

For example, if I_{INFS} is 20mA and V_{INFS} is to be 2V, $R_C = 100\Omega$ and $R_G \cong 1.19k\Omega$, which might consist of a 1.13kΩ resistor and a 100Ω trim potentiometer.

If the input current is zero-based, e.g., 0 to 50mA, (15.2) becomes

$$R_G = \frac{6.31 V_{INFS}}{10.1 - V_{INFS}} k\Omega \tag{15.3}$$

It is also useful to convert a 4-to-20mA current to frequency for noise immunity, isolation, and ease of going digital. If the output frequency is to be scaled directly (say 1kHz to 5kHz), the circuitry, using IC V/f converters, such as the AD537, is straightforward. Isolation can be combined with self-powering from the

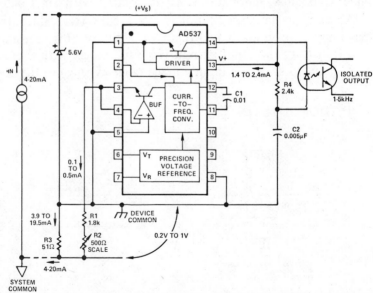

Figure 15-2. Isolated 4-20mA-to-frequency converter

loop current, as Figure 15-2 shows. All required power is provided by a portion of the minimum loop current of 4mA, flowing through the AD537's internal circuitry, with the excess flowing through the 5.6V Zener diode, which establishes the supply voltage. As will be shown, the output is optically coupled.

Note that there are two return paths for the signal current: through the main load resistor, R3, and through the timing resistance, $(R_1 + R_2)$. Since the timing current is theoretically proportional to the load current in R3, linearity is preserved (even though the current gets used in a variety of ways by the rest of the circuit).

The supply voltage, provided by the Zener diode, together with the 1V full-scale voltage across R3, adds up to a worst-case compliance of 7V.

Each cycle, capacitor C2 is discharged through the LED at a peak current of approximately 25mA, internally limited in the AD537. The value of C2 is chosen to produce an optical pulse of about 1μs duration. The maximum average current through the diode is limited to about 2mA by R4, which is calculated to insure that, even when the signal current is at its minimum value of 4mA, there is at least 0.5mA flowing through the Zener; thus the optical output persists at a lower level for a half-cycle. During the next half-cycle, C2 recharges via R4, with a time-constant of 12μs, allowing full recharging in the minimum half-cycle time of 100μs.

In many cases, the only calibration required will be to set the frequency to 5kHz at the full-scale input of 20mA. However, the V_{OS} of the AD537 may also be nulled for more exacting applications, using standard techniques.[1]

MORE THOUGHTS ON ISOLATION

One of the most important considerations about using an isolation amplifier is the manner in which it is hooked up. The following guidelines, if observed, will help a user to realize the full performance capability of the isolator and minimize spurious noise and pickup. Since the more-common sources of electrical noise arise from ground loops, electrostatic coupling and electromagnetic pickup (Chapter Three), these guidelines concern the guarding of low-level millivolt signals in hostile environments.

---

[1] For more information, see the Application Note: "Applications of the AD537 IC Voltage-to-Frequency Converter," by Doug Grant, available from Analog Devices.

● Use twisted shielded cable to reduce inductive and capacitive pickup.

● Drive the transducer cable shield, S, with the common-mode signal source, E_G, where possible, to reduce the effective cable capacitance, as Figure 15-3 shows. This is accomplished by connecting to the signal low point, B. However, this may not always be possible. In some cases, for example, the shield must be separated from signal low by a portion of the medium being measured, causing a common-mode signal, E_M, to appear between the shield and signal low. The CMR capability between the input terminals (HI IN and LO IN) and GUARD will work to suppress that common-mode signal, E_M.

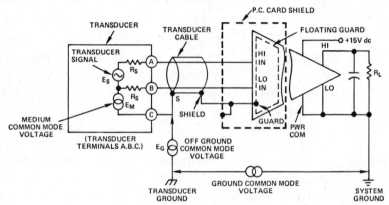

Figure 15-3. Transducer-amplifier interconnection

● To avoid ground loops and excessive hum, signal low, B, or the transducer cable shield, S, should never be grounded at more than one point.

● Dress unshielded leads short at the connection terminals and reduce the area formed by these leads to minimize inductive pickup.

Multi-channel isolators have some uses that have not been discussed elsewhere in this volume. Two of them are depicted in Figures 15-4 and 15-5.

In Figure 15-4, synchronized isolators (288K) and their driver (947) are shown as the three-channel front end of a data-acquisition system which provides up to 850V of both input-to-output isolation and channel-to-channel input isolation. Thus, the transducers can all float; ground-loop problems between the transducers, and ground loops between the inputs and output ground are elimi-

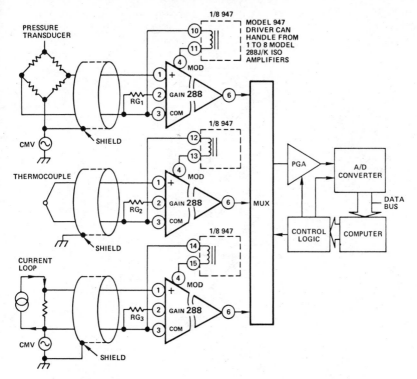

Figure 15-4. 3-channel isolated data-acquisition system front end. The 2B54 isolated multiplexer could also be used in this application. See Figure 15-14.

nated. The high CMV rating and input protection provide protection for the multiplexer against transients, fostering reliable operation in hostile environments. Other characteristics that help maintain the integrity of the data are the 288K's high CMR (92dB) and low nonlinearity (0.05%), all at low cost in multichannel applications.

In Figure 15-5, a number of strain, pressure, or temperature-measurement bridges are driven by current in a series string, to ensure that they receive uniform excitation (if their resistances are uniform), with only a single supply.

Naturally, the current-forcing scheme results in appreciable CMV levels, especially for the transducers at the high end of the chain, so an amplifier with both high tolerance of CMV and high common-mode rejection is required.

Since this application calls for a multi-channel, high-accuracy isolator with only moderately high CMV protection (less than

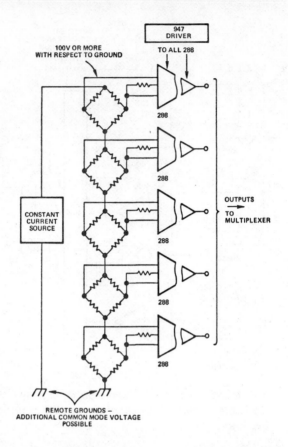

Figure 15-5. Isolation amplifiers in high-CMV, current-excited bridge scheme. Model 289, 3-port synchronizable isolators could also be used here.

850V), 288's, with their 947 driver, are an appropriate approach.

TOPICS IN FILTERING

Most of the filters dealt with in Chapter Three have fixed time constants. Variable time constants are often useful, as the following examples show.

In Figure 13-2, a voltage-controlled low-pass filter made it possible to remotely change the filtering time constant to either filter out periodic pulsations or to observe rapid changes when necessary. The same kind of filter is shown in Figure 15-6, for a somewhat more-detailed discussion. The output voltage of the multiplier is self adjusted by the feedback loop to maintain the voltage across R proportional to the product of the control input and the signal-input-less-the-capacitor-voltage. Since the capacitor voltage is proportional to the integral of the current, the response at A (with a high-impedance load) is that of a unit-lag filter with cutoff frequency proportional to E_C. In addition to that response,

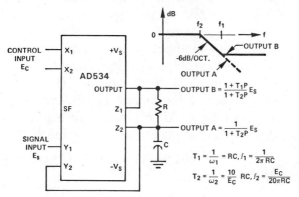

Figure 15-6. Voltage-controlled low-pass filter

output B provides a lag-lead characteristic (variable lag, fixed lead).

Output B is at low impedance and is essentially load-independent, but the voltage at output A should be buffered if its destination is not at high impedance.

Filters can be programmed digitally, too. Figure 15-7 shows an application of a multiplying d/a converter as a digitally controlled first-order low-pass filter. Cutoff frequency is proportional to D, the fractional value of the digital input $(0 \leqslant D \leqslant (1 - 2^{-n}))$, for fractional binary). Frequency scale factor can be adjusted pro-

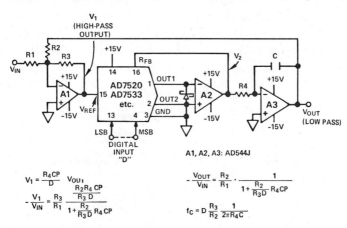

Figure 15-7. Digitally variable first-order filter

portionally with R3, time constant can be trimmed proportionally with R4, overall gain can be trimmed independently with R1.

This circuit and the one immediately preceding it both use integrators in feedback loops. As the programmed cutoff frequency is reduced, the loop gain is correspondingly reduced at low frequencies, in the same way that it is when the denominator is reduced in an analog multiplier used as a divider. Therefore, it is important to use adequately large time constants in the integrator (RC in 15-6, R_4C in 15-7), so that E_C and D approach full-scale as closely as possible at the highest cutoff frequency. For large dynamic ranges of E_C or D, it is useful to trim circuit offset to zero with E_C or D held at the low end of the range.

In Figure 15-7, the output of A1 and A2 are proportional to the derivative of the output,

$$-\frac{V_1}{V_{IN}} = \frac{R_3}{R_1} \cdot \frac{Tp}{1+Tp} = \frac{R_3}{R_1} \frac{j\frac{\omega}{\omega_C}}{1+j\frac{\omega}{\omega_C}} \tag{15.4}$$

where $T = \frac{R_2R_4C}{R_3D}$ and $\omega_C = \frac{1}{T} = \frac{R_3D}{R_2R_4C}$; $f_c = \frac{R_3D}{2\pi R_2R_4C}$.

(15.4) will be recognized as a first-order high-pass filter response, with the cutoff frequency proportional to the digital input. R_3/R_1 should be the largest value that will not cause saturation of A1 or A2 when the input is a step function (or the most-rapid change expected).

Figure 15-8 shows a simple way of changing the time-constant of a low-pass filter from one value to another. In a similar approach to that used for the scale in Figure 12-5, this circuit permits rapid response to a step function, then a longer time-constant for filtering high-frequency noise. When the switch is closed, the time-constant

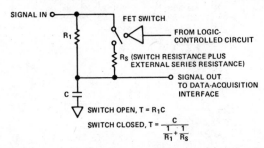

Figure 15-8. Low-pass filter with switched time constant

is short; when the switch opens, the time-constant increases. For applications where it is suitable, such as in batch processes (e.g., weighing discrete increments of load), it combines the benefits of rapid response and good filtering action.

Figures 15-9 and 15-10 depict nonlinear filters that achieve an effect similar to that of Figure 15-8, but in response to the signal itself, and with a smooth transition. They may be termed *derivative-controlled low-pass filters*.

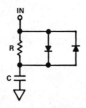

Figure 15-9. Simple derivative-controlled low-pass filter

Figure 15-9 shows how simple such a filter can be. It consists of an RC, with the resistor paralleled by a pair of diodes. If the input is a slowly varying signal with small fast noise signals riding on it, the output will be determined by the filter's RC. However, if the input signal changes rapidly by more than one diode drop (about 0.6V), the diodes will go to low impedance and bypass the resistor, providing rapid response. However, as the capacitor charges and the voltage drop across the diodes decreases, their resistance increases in approximately logarithmic fashion, more of the current flows through the resistor, and the response will increasingly depend on RC.

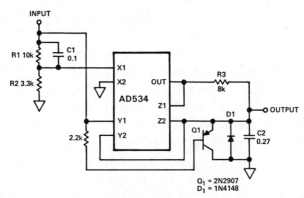

Figure 15-10. Circuit of computing-type derivative-controlled low pass filter

Figure 15-10 is a circuit having similar effect but more rapid and predictable response (and a greater number of components). It will be recognized basically as the voltage-controlled filter of Figure 15-6. The signal input goes directly to Y1. It also goes through a differentiating network, R1, R2, C1, and this *lead* signal modulates the filter time constant.

When a step voltage is applied to the input, the time constant is immediately determined by the full value of the step (which appears at X1); then the control voltage exponentially retracts until it is about 25% of the step (the divider ratio $R_2/(R_1 + R_2)$), which increases the filter time-constant fourfold.

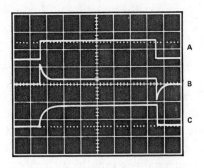

Figure 15-11. Responses of filter of Figure 15-10

Figure 15-11 shows salient waveforms in the circuit. Trace A is the input step. Trace B is the control input, showing the immediate jump in cutoff frequency, and the decay to a cutoff frequency about 1/4 as large. Trace C is the signal output, showing a rapid response, followed by a long tail, indicating the low steady-state cutoff frequency.

When the square-wave returns, the spike applied to X1 will go in the wrong direction, tending to greatly lengthen the time constant, unless a clamping circuit (consisting of Q1 and D1) is used to rapidly reset C2's charge to zero. If derivative control of the falling edge were desired, an absolute-value amplifier could be inserted between the X1 input and the R1-R2 junction.

In this example, the steady-state time constant is determined by the height of the step and/or the ratio of the resistances R_1/R_2. If there are to be other influences on the time constant, appropriate additional voltages (of opposite polarity) could be summed in at the X2 input. This circuit will function properly only if the noise is relatively small and is not pulse-or step-type. As noted earlier, this approach to filtering has proven useful in electronic weighing applications, where long time-constants are undesirable when weighing an object, yet floor-noise has to be filtered out.

PROGRAMMABLE-GAIN ISOLATOR

Gain, too, can be switch programmed. For example, in wide-dynamic range bridge-type measurements, where gain-ranging would be a desirable feature of the system instrumentation, an isolation amplifier can provide both bridge excitation and signal conditioning. In the application shown in Figure 15-12, the un-committed op-amp front end and isolated dual 15V supplies of the 277 Isolator permit gain to be set remotely at a value that optimizes input signal-to-noise ratio and eliminates the need for high-quality post-amplifiers at the isolator output.

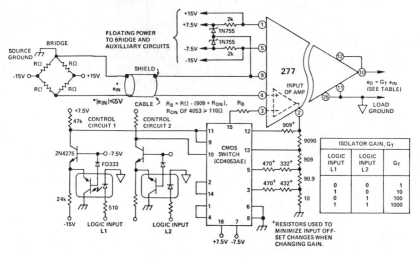

Figure 15-12. Programmable-gain bridge transducer amplifier

Control switches are driven by TTL inputs, which are isolated from, source ground by the opto-isolators in the control switch. Control signals operate the CMOS switch network to establish the gains listed in the table. The CMOS switch network is operated in a manner that causes the resistance of the switches to be in series only with the negative input of the isolator, and not in series with the gain-setting resistors; therefore the resistance of the CMOS switches does not affect the gain. Resistor R_B is connected in series with the negative input to reduce errors due to bias-current drift. (If R_B calculates negative, the resistor should be connected in series with the + input.)

HIGH-PERFORMANCE FLOATING DATA AMPLIFIER

The amplifier configuration shown in Figure 15-13 is a chopper-stabilized isolated amplifier capable of withstanding 2500V of continuous common-mode voltage with dc CMR of 160dB (100,000,000:1), and having dc drifts better than 100nV/°C. Its gain of 200,000 permits microvolts to be measured, and volts to be delivered at the output.

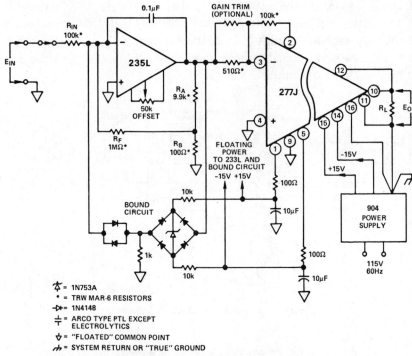

Figure 15-13. Floating input data amplifier

This high-performance amplifier is created by combining the low drift and excellent isolation characteristics of the 277 isolation amplifier with the dc stability of a 235L chopper-stabilized operational amplifier. The 235 is powered by the 277's floating supply and serves as a fully floating gain-of-1000 preamplifier. The output of the 277 is an isolated, amplified version of the 235's output. A five-microvolt signal unbalance at the 235's input will yield one volt at the 277's output.

In this example, the 235 is connected as a gain-of-1000 inverting amplifier, with an input impedance of 100kΩ. While this is certainly acceptable for most applications, a non-inverting chopper (e.g., Model 261) instead of the 235 will provide very high input impedance for applications that require it, with perhaps a small sacrifice in bandwidth. High gain is achieved by feeding 1/100th of the 235's output back to the input via a 1MΩ resistance (additional gain-ratio of 10). This produces high gain without resorting to high values of feedback resistance, which invite leakage and stability problems.

Gain adjustment may be done at the 100Ω resistor, R_G. This eliminates any possible contribution to error at the summing point and allows an inexpensive switch to be used.

The menagerie of diodes which parallel the feedback path act as a bound on the output swing of the 235 in order to ensure fast recovery from input overload. At a gain of 100, only *12 millivolts* of signal is needed to slam the amplifier into saturation. Chopper-stabilized amplifiers can require significant time to recover to zero offset after an input overload has occurred; the bound circuit lowers the gain of the amplifier, at high output levels, keeping its output from saturating and speeding recovery.

The bandwidth is deliberately limited to a few Hz by the 0.1μF capacitor, to minimize high-frequency noise transmission. For most measurements of the types discussed in this book, such response is adequate.

The drift characteristics of chopper-stabilized amplifiers are insensitive to input offset adjustments. For this reason, the offset adjustment of the 235 is used to zero the entire amplifier. If gain accuracy is important, the 510Ω resistor can be parallel-trimmed to achieve a precise value of overall gain. Since the 277 is connected as an inverting amplifier, with a gain of 200, its output will be of the same polarity as the input signal (unless the 261 non-inverting amplifier is used as the preamp).

Each amplifier should be enclosed in a "house" constructed of copper walls. The amplifiers (especially the 235) should be packed in Styrofoam or Fiberglas if thermal transients are present. The copper boxes offer protection from stray fields, from RF and static pickup, and from unwelcome conversations between the carrier circuits of the 277 and the 235. (Any coherence between the two frequencies or their harmonics can result in beat frequencies, which masquerade as dc uncertainty.)

If this seems like a great deal of trouble to go to in order to construct an amplifier, it is—but it's worth the effort. This circuit, properly embodied with respect to grounding, shielding, component choice, and good wiring practice, has yielded a powerful low-offset-error dc amplifier—higher in performance than any instrumentation amplifier commercially available to date.

Salient performance that has been achieved includes:

Gain: 200,000 (higher gains readily achievable)
Drift: 100nV/°C (typically 50nV/°C, 20° to 30°C
Time drift: 5µV/year
CMRR (dc):   160dB (100,000,000:1)
Common-mode voltage: 2500V continuous
Input noise: 1µV, peak-to-peak, in a 1Hz bandwidth

This kind of offset and common-mode performance will find use in transducer signal-conditioning where the highest grade of performance is required. The sole errors of consequence in this design are the 0.05% nonlinearity and the 50ppm/°C gain drift of the 277J, which—though quite low—should be considered if necessary.

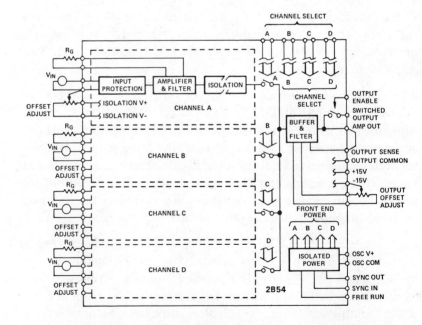

Figure 15-14. Isolated 4-channel multiplexer (2B55)

ISOLATED ALL-ELECTRONIC MULTIPLEXING

In Chapter Seven, there was a discussion of the role of a low-level multiplexer (the 2B54) in thermocouple measurement (see Figure 7-8). Figure 15-14 is a somewhat more detailed block diagram of the 2B54 and a companion high-level multiplexer, the 2B55.

As the diagram shows, there are four channels, isolated from one another, from the output, from the decoded channel-select inputs, and from the power supply for up to 750V rms. Each channel is also protected against 130V rms ac differential input voltage and has individual gain and offset adjustments, and an internal filter. Open inputs can be detected, and the output is protected against continuous short circuits to either supply or ground.

The four channels share a common oscillator, which can be synchronized to the oscillators of additional multiplexers if more than four channels are desired. Because a "three-state analog" output connection is available, additional ranks of multiplexing are not required; the outputs can be connected in common and enabled individually.

A safe, active, solid-state, *all-electronic* multiplexer, the 2B54 (2B55) is a superior alternative in many applications that heretofore have been restricted to reed-relay and flying-capacitor techniques.

PULSED-MODE BRIDGE EXCITATION

The first figure in this book concerns itself with a method for pulsing large inputs to an active transducer to obtain proportionally increased output voltages without excessive heating of the element.

Figure 15-5 is a more-informative version of the principle embodied in Figure 1-1. A bridge is suspended (electrically) between the collectors of a PNP-NPN switch referred to ±50V. For this reason, the common-mode voltage is close to zero volts, and the requirements for amplifier common-mode rejection are not severe. If the bridge were switched from ground or supply only, common-mode potentials of 50V would result, and an isolation amplifier would be required.

Although more intricate than the usual bridge-conditioning circuit, this technique is a useful way to reduce the effect of the monitoring amplifier's drift. With 100 volts being switched, instead of the usual steady ten volts, an AD521's effective drift in

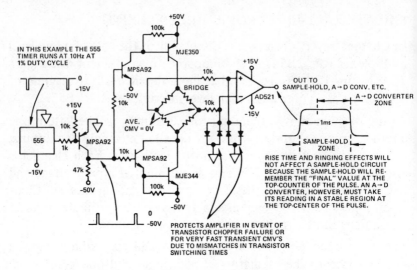

Figure 15-15. Pulsed high-voltage-bridge drive

the former case is $0.5\mu V/°C$, while the low-drift AD522 would have an effective drift of $0.2\mu V/°C$, and the 234 chopper-stabilized amplifier would have, in effect, $10nV/°C$!

Naturally, failures in the transistor drive-circuitry could cause disastrous consequences to the bridge circuit and the amplifier. The opto-isolator and associated components in Figure 15-16 are intended to cause the SCR crowbar to trigger and blow the fuse in the event of such a failure.

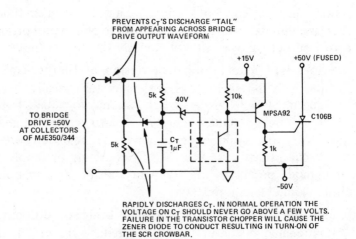

Figure 15-16. Protective circuit for high-voltage bridge

The duty cycle and the *on* time of the transistor chopper must be set with an eye towards the thermal limitations of the bridge elements, the settling-time-to-desired-accuracy of the amplifier amployed, and the characteristics of the sample-hold used to read the output of the amplifier.

APPENDIX

ACCURACIES OF THE AD590

Maximum errors over limited temperature spans, with $V_S = +5V$, are listed by device grade in the following tables. The tables reflect the worst-case departures of the AD590 from linearity; they invariably occur at the extremities of the specified temperature range. The trimming conditions for the data in the tables are shown in Figures A-1 and A-2. The data is subject to the considerations in the Notes, which follow the tables.

All errors listed in the tables are $\pm°C$. For example, if $\pm1°C$ maximum error is required over the $+25°C$ to $+75°C$ range (i.e., lowest temperature of $+25°C$ and span of $50°C$), then the trimming of a J-grade device, using the single-trim circuit (Figure A-1), will result in output having the required accuracy over the stated range. An M-grade device with no trims will have less than $\pm0.9°$ error, and an I-grade device with two trims (Figure A-2) will have less than $\pm0.2°$ error. If the requirement is for less than $\pm1.4°C$ max error, from $-25°C$ to $+75°C$ ($100°$ span from $-25°C$), it can be satisfied by an M-grade device with no trims, a K-grade device with one trim, or an I-grade device with two trims.

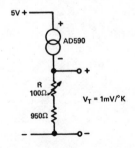

Figure A-1a. Single-trim calibration circuit

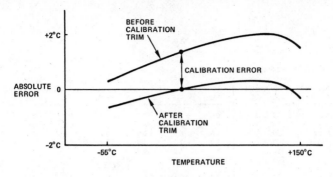

b. *Effect of single calibration trim on accuracy*

Figure A-1. One-temperature calibration trim

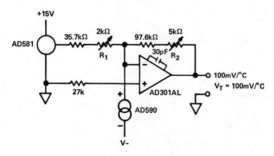

a. *Two-temperature trim circuit*

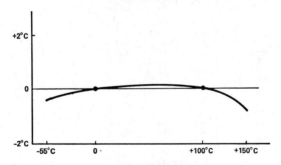

b. *Typical two-trim accuracy*

Figure A-2. Two-temperature offset and gain trim

M GRADE — Maximum Errors, °C

Number Of Trims	Temperature Span — °C	Lowest Temperature In Span — °C							
		-55	-25	0	+25	+50	+75	+100	+125
None	10	0.6	0.5	0.6	0.6	0.7	0.7	0.7	0.9
None	25	0.8	0.8	0.7	0.7	0.8	0.8	1.0	1.1
None	50	1.0	0.9	0.8	0.9	0.9	1.1	1.2	
None	100	1.3	1.4	1.3	1.4	1.5			
None	150	1.5	1.6	1.6					
None	205	1.7							
One	10	0.2	0.1	0.1	0.1	0.1	0.1	0.1	0.2
One	25	0.4	0.3	0.2	0.2	0.2	0.2	0.3	0.4
One	50	0.5	0.4	0.3	0.3	0.3	0.4	0.5	
One	100	0.8	0.8	0.7	0.7	0.8			
One	150	0.9	0.9	0.9					
One	205	1.0							
Two	10	0.1	<	<	<	<	<	<	0.1
Two	25	0.1	<	<	<	<	<	<	0.1
Two	50	0.2	<	<	<	<	<	0.2	
Two	100	0.2	0.1	<	0.1	0.2			
Two	150	0.3	0.2	0.3					
Two	205	0.3							

<: Less than ±0.05°C

L GRADE — Maximum Errors, °C

Number Of Trims	Temperature Span — °C	Lowest Temperature In Span — °C							
		-55	-25	0	+25	+50	+75	+100	+125
None	10	1.0	1.0	1.1	1.1	1.2	1.3	1.4	1.6
None	25	1.3	1.3	1.3	1.4	1.5	1.6	1.7	1.9
None	50	1.9	1.8	1.7	1.8	1.9	2.1	2.4	
None	100	2.4	2.4	2.4	2.4	2.7			
None	150	2.7	2.6	2.8					
None	205	3.0							
One	10	0.2	0.1	0.1	0.1	0.1	0.1	0.1	0.2
One	25	0.5	0.4	0.3	0.3	0.3	0.3	0.4	0.5
One	50	1.0	0.8	0.6	0.6	0.6	0.8	1.0	
One	100	1.3	1.2	1.1	1.1	1.3			
One	150	1.4	1.3	1.4					
One	205	1.6							
Two	10	0.1	<	<	<	<	<	<	0.1
Two	25	0.1	<	<	<	<	<	<	0.1
Two	50	0.2	<	<	<	<	<	0.2	
Two	100	0.3	0.2	0.1	0.2	0.3			
Two	150	0.3	0.2	0.3					
Two	205	0.4							

<: Less than ±0.05°C

K GRADE — Maximum Errors, °C

Number Of Trims	Temperature Span — °C	Lowest Temperature in Span — °C							
		-55	-25	0	+25	+50	+75	+100	+125
None	10	2.1	2.3	2.5	2.7	2.9	3.1	3.3	3.6
None	25	2.6	2.7	2.8	3.0	3.2	3.5	3.8	4.2
None	50	3.8	3.5	3.4	3.6	3.8	4.3	5.1	
None	100	4.2	4.3	4.4	4.6	5.1			
None	150	4.8	4.8	5.3					
None	205	5.5							
One	10	0.2	0.1	0.1	0.1	0.1	0.1	0.1	0.2
One	25	0.6	0.4	0.3	0.3	0.3	0.4	0.5	0.6
One	50	1.2	1.0	0.7	0.7	0.7	1.0	1.2	
One	100	1.5	1.4	1.3	1.3	1.5			
One	150	1.7	1.5	1.7					
One	205	2.0							
Two	10	0.1	<	<	<	<	<	<	0.1
Two	25	0.2	0.1	<	<	<	<	0.1	0.2
Two	50	0.3	0.1	<	<	<	0.1	0.2	
Two	100	0.5	0.3	0.2	0.3	0.7			
Two	150	0.6	0.5	0.7					
Two	205	0.8							

<: Less than ±0.05°C

J GRADE — Maximum Errors, °C

Number Of Trims	Temperature Span — °C	Lowest Temperature In Span — °C							
		-55	-25	0	+25	+50	+75	+100	+125
None	10	4.2	4.6	5.0	5.4	5.8	6.2	6.6	7.2
None	25	5.0	5.2	5.5	5.9	6.0	6.9	7.5	8.0
None	50	6.5	6.5	6.4	6.9	7.3	8.2	9.0	
None	100	7.7	8.0	8.3	8.7	9.4			
None	150	9.2	9.5	9.6					
None	205	10.0							
One	10	0.3	0.2	0.2	0.2	0.2	0.2	0.2	0.3
One	25	0.9	0.6	0.5	0.5	0.5	0.6	0.8	0.9
One	50	1.9	1.5	1.0	1.0	1.0	1.5	1.9	
One	100	2.3	2.2	2.0	2.0	2.3			
One	150	2.5	2.4	2.5					
One	205	3.0							
Two	10	0.1	<	<	<	<	<	<	0.1
Two	25	0.2	0.1	<	<	<	<	0.1	0.2
Two	50	0.4	0.2	0.1	<	<	0.1	0.2	<
Two	100	0.7	0.5	0.3	0.7	1.0			
Two	150	1.0	0.7	1.2					
Two	205	1.5							

<: Less than ±0.05°C

I GRADE — Maximum Errors, °C

Number Of Trims	Temperature Span — °C	Lowest Temperature In Span — °C							
		-55	-25	0	+25	+50	+75	+100	+125
None	10	8.4	9.2	10.0	10.8	11.6	12.4	13.2	14.4
None	25	10.0	10.4	11.0	11.8	12.0	13.8	15.0	16.0
None	50	13.0	13.0	12.8	13.8	14.6	16.4	18.0	
None	100	15.2	16.0	16.6	17.4	18.8			
None	150	18.4	19.0	19.2					
None	205	20.0							
One	10	0.6	0.4	0.4	0.4	0.4	0.4	0.4	0.6
One	25	1.8	1.2	1.0	1.0	1.0	1.2	1.6	1.8
One	50	3.8	3.0	2.0	2.0	2.0	3.0	3.8	
One	100	4.8	4.5	4.2	4.2	5.0			
One	150	5.5	4.8	5.5					
One	205	5.8							
Two	10	0.3	0.2	0.1	<	<	0.1	0.2	0.3
Two	25	0.5	0.3	0.2	<	0.1	0.2	0.3	0.5
Two	50	1.2	0.6	0.4	0.2	0.2	0.3	0.7	
Two	100	1.8	1.4	1.0	2.0	2.5			
Two	150	2.6	2.0	2.8					
Two	205	3.0							

<: Less than ±0.05°C

NOTES

1. Accuracy over the 205°C span is tested for all devices by a 4-point measurement, at –55°C, +25°C, +125°C, and +150°C. Maximum errors over other ranges are guaranteed (but not tested), based on the known characteristic behavior of the AD590.

2. For one-trim accuracy specifications, the 205°C span is assumed to be trimmed at +25°C; for all other spans, it is assumed that the device is trimmed at the midpoint.

3. For the 205°C span, it is assumed that the two trim temperatures are in the vicinity of 0°C and +140°C; for all other spans, the specified trims are at the endpoints.

4. In precision applications, the actual errors encountered are usually dependent upon sources of error which are often overlooked in error budgets. These typically include:
   a. Trim error in the calibration technique used
   b. Repeatability error
   c. Long-term drift errors

Trim error is usually the largest error source. This error arises from such causes as poor thermal coupling between the device to be calibrated and the reference sensor; reference sensor errors; lack of adequate time for the device being calibrated to settle to the final temperature; radically different thermal resistances between the case and the surroundings (θ_{CA}) when trimming and when applying the device.

Repeatability errors arise from a strain hysteresis of the package. The magnitude of this error is solely a function of the magnitude of the temperature span over which the device is used. For example, thermal shocks between 0°C and 100°C involve extremely low hysteresis and result in repeatability errors of less than ±0.05°C. When the thermal-shock excursion is widened to −55°C-to-+150°C, the device will typically exhibit a repeatability error of ±0.05°C (±0.10 guaranteed maximum).

Long-term drift errors are related to the average operating temperature and the magnitude of the thermal shocks experienced by the device. Extended use of the AD590 at temperatures above 100°C typically results in long-term drift of ±0.03°C per month; the guaranteed maximum is ±0.10°C/mo. Continuous operation at temperature below 100°C induces no measurable drifts in the device. Besides the effects of operating temperature, the severity of thermal shocks incurred will also affect absolute stability. For thermal-shock excursions less than 100°C, the drift is difficult to measure (<0.03°C). However, for 200°C excursions, the device may drift by as much as ±0.10°C after twenty such shocks. If severe, quick shocks are necessary in the application of the device, realistic simulated life tests are recommended for a thorough evaluation of the error introduced by such shocks.

BIBLIOGRAPHY

This Bibliography is intended to provide a sampling of publications likely to prove useful to the reader. In many cases, they fan out to additional refences, for readers seeking information in greater depth. *Only* those publications marked with an asterisk (*) are available from Analog Devices (at no charge).

The bibliography is grouped into five classes: Publications of Analog Devices; examples of manufacturers' literature; trade books; society publications, directories, serials, and other compendia; magazine articles.

PUBLICATIONS OF ANALOG DEVICES, INC.

Analog Dialogue. *
    Published several times per year (ISSN: 0161-2636). Timely technical information on Analog Devices technologies, products, and their applications.

Analog Productlog.
    Published quarterly. Timely descriptions of new Analog Devices products for precision measurement and control.

Application Guide for CMOS Multiplying D/A Converters. Analog Devices, 1978*

Boyes, G., *ed. Synchro and Resolver Conversion.* 208 pp. Norwood, MA 02062: Analog Devices, 1980, Published in the U.K., ISBN 0-916550-06-0, $11.50.

Brokaw, A. Paul. "Analog Signal Handling for High Speed and Accuracy." *Analog Dialogue* 11-2, 1977.*
    Grounding, bypassing, compensation, and avoiding offset-induced nonlinearity in d/a-converter circuitry.

Brokaw, A. Paul. "An I.C. Amplifier Users' Guide to Decoupling, Grounding, and Making Things Go Right for a Change." Application Note. Analog Devices, reprinted, 1979*.
    The title is self-explanatory.

Burton, D. P., Dexter, A. L. *Microprocessor Systems Handbook.* Norwood, MA 02062: Analog Devices, 1977, ISBN 0-916550-04-4, $9.50.

Data Acquisition Products Catalog. Analog Devices, 1978 (or latest available edition).*
    Tutorial information, selection guides, and complete data sheets on amplifiers, other signal-conditioning products, and conversion products.

Data Acquisition Products Catalog Supplement. Analog Devices, 1979 (or latest available edition).*
    Selection guides and complete data sheets issued since the last *Catalog.*

Digital Panel Instrument Catalog Analog Devices, 1979 (or latest available edition).*
    Orientation, selection guide, and data sheets on DPMs, temperature-measurement, and signal-conditioning components.

Fishbeck, James. "Writing P-I-D Control Loops Easily in BASIC." *Control Engineering*, October, 1978, pp. 45-47.

Grant, Doug. "Applications of the AD537 IC Voltage-to-Frequency Converter." Application Note. Analog Devices, 1978.*

Integrated Circuit Chips for Precision Hybrids. Analog Devices, 1979.*

Chip catalog: Data sheets of CMOS and bipolar data-conversion and signal-conditioning IC device chips with guaranteed specifications.

Isolation and Instrumentation Amplifier Designers' Guide. Analog Devices, 1979.*

Kress, David. "Monolithic Sample/Hold Amplifiers: IC flexibility comes to the sample/hold scene." *Electronic Products Magazine*, March, 1977, pp. 41-46.*

Multiplier Application Guide. Analog Devices, 1978.*

Basics, applications, theory, and technical data on analog multipliers and dividers.

Product data sheets. Analog Devices.*

Data sheets are available individually for all products, with 2-16 pages, as necessary. They include features, descriptions, specifications, and application information. From time to time, data sheets are collected and published in catalog form.

Riskin, Jeffrey R. "A User's Guide to IC Instrumentation Amplifiers." Application Note. Analog Devices, 1978.*

Sheingold, Daniel H., ed. *Analog-Digital Conversion Notes*. Norwood, MA 02062; Analog Devices, Inc., 1977, ISBN 0-916550-03-6, $5.95

Sheingold, Daniel H., ed. *Nonlinear Circuits Handbook: Designing with analog function modules and IC's*. Norwood, MA 02062: Analog Devices, Inc., 1974, ISBN 0-916550-01-X, $5.95.

1980 (or most recent) *Short-Form Guide: Electronic products for precision measurement and control*. Analog Devices.*

A complete listing of currently available products.

Synchro-to-Digital Converters. Short-form guide. Analog Devices, 1979 (or most recent edition).*

EXAMPLES OF USEFUL MANUFACTURERS' LITERATURE

The Application of Filters to Analog and Digital Signal Processing. West Nyack, N.Y.: Rockland Systems Corporation, 1976.

Applications Manual for Operational Amplifiers, Teledyne Philbrick, 1968.

Brown, James A. "Temperature Measurement and Control." Boonton, NJ: RFL Industries, Inc., 1971.

The CAMBION Thermoelectric Handbook. Cambridge, MA: Cambridge Thermionic Corporation, 1972.

The Capsule Thermistor Course. Framingham, MA: Fenwal Electronics, 1979.

Electrometer Measurements. Cleveland: Keithley Instruments, 1977. Good information on low-level measurement techniques.

Elimination of Noise in Low-Level Circuits. Cleveland: Gould Inc., Instrument Systems Division.

Fluke. "Guide to Temperature Measurement." *Instruments & Control Systems,* Radnor, PA: Chilton Co., June, 1979, 4-color wallchart insert.

Handbook of Measurement and Control. Camden, NJ: Schaevitz Engineering, 1976.

Hueckel, John H. "Input Connection Practices for Differential Amplifiers." Duarte, CA: Neff Instrument Corporation.

NANMAC Temperature Handbook 1979/80. Framingham, MA: NANMAC Corporation, 1979.

OMEGA Temperature Measurement Handbook. Stamford, Conn.: OMEGA Engineering, 1980 (published annually).

"Pressure Transducer Application Notes," 1–7, loose pages. Freeport, IL: Microswitch, 1978 and 1979.

The Pressure Transducer Handbook. Santa Clara, CA: National Semiconductor, 1977.

Signal Conditioning. Cleveland, Ohio: Gould, Inc., Brush Instruments Division, November, 1969.

SR-4 Strain Gage Handbook. Waltham, MA: BLH Electronics, 1980.

Thermistor Catalog. Yellow Springs, Ohio: Yellow Springs Instrument Company.

Thermoelectric Thermometry. Mountlake Terrace, WA: John Fluke Mfg. Co. 13-page pamphlet full of good practical info.

BOOKS

Benedict, Robert P. *Fundamentals of Temperature, Pressure, and Flow Measurements.* New York: John Wiley & Sons, 1977.

Considine, Douglas M., ed. *Process Instruments and Controls Handbook.* 2d ed. New York: McGraw-Hill Book Company, 1974.

Clayton, George B. *Linear Integrated Circuit Applications.* London: The MacMillan Press Ltd., 1975.

Coughlin, Robert F., Driscoll, Frederick F. *Operational Amplifiers and Linear Integrated Circuits.* Englewood Cliffs, NJ: Prentice-Hall, 1977.

Herceg, Edward E. *Handbook of Measurement and Control: An authoritative treatise on the theory and application of the LVDT.* HB-76. Pennsauken, NJ: Schaevitz Engineering, 1976.

Hoenig, Stuart A., Payne, F. Leland. *How to Build and Use Electronic Devices without Frustration, Panic, Mountains of Money, or an Engineering*

Degree. Boston: Little Brown and Company, 1973.

Johnson, Curtis D. *Process Control Instrumentation Technology.* New York: John Wiley & Sons, 1977.

 Good basic practical textbook.

Jung, Walter G. *IC Op-Amp Cookbook.* Indianapolis: Howard W. Sams & Co., Inc., 1974 (or latest edition).

Kaufman, Milton and Seidman, Arthur H., eds. *Handbook of Electronic Calculations.* New York: McGraw-Hill, 1979.

Kinzie, R. A. *Thermocouple Temperature Measurement.* New York: Wiley, 1973.

Lion, Kurt. *Instrumentation in Scientific Research.* New York: McGraw-Hill, 1959.

Morrison, Ralph. *Grounding and Shielding Techniques in Instrumentation.* 2d. ed. New York: Wiley Interscience, 1977.

A *must* for any serious circuit and system designer.

Norton, Harry N. *Handbook of Transducers for Electronic Measuring Systems.* Englewood Cliffs, NJ: Prentice-Hall, 1969.

Ott, Henry W. *Noise Reduction Techniques in Electronic Systems.* New York: Wiley-Interscience, 1976.

Roberge, James K. *Operational Amplifiers, Theory & Practice.* New York: John Wiley & Sons, Inc., 1975.

Sachse, H. B. *(Semiconducting Temperature Sensors and Their Applications.* New York: Wiley, 1975.

Smith, John I. *Modern Operational Circuit Design.* New York: Wiley-Interscience, 1971.

Soisson, H. E. *Instrumentation in Industry.* New York: Wiley, 1975.

Stout, David F., Kaufman, Milton, ed. *Handbook of Operational Amplifier Circuit Design.* New York: McGraw-Hill Book Company, 1976.

Tobey, G. E., Graeme, J. G., and Huelsman, L. P. *Operational Amplifiers: Design and Applications.* New York: McGraw-Hill, 1971.

SOCIETY PUBLICATIONS, DIRECTORIES, SERIALS, ETC.

CEM: Chilton's Control Equipment Master '79-'80. Radnor, PA: Chilton Company, 1979 (or most-recent).

"Compatibility of Analog Signals for Electronic Industrial Process Instruments." ISA Standard S50.1 Pittsburgh: Instrument Society of America, 1975.

Edelman, Sheldon, "Glossary of Microprocessor-Based Control System Terms." *Instruments & Control Systems,* May, 1979, pp. 43–48.

EDN. 221 Columbus Ave., Boston, MA 02116, Tel. (617) 536-7780.

Electronic Design. 50 Essex St., Rochelle Park, NJ 07662, Tel. (201) 843-0550.

Electronic Design's GOLD BOOK. 2 volumes. Rochelle Park NJ: Hayden Publishing Co., published bi-annually.

Electronic Engineers Master Catalog. 2 volumes. Garden City, NY: United Technical Publications, published bi-annually.

Electronic Products Magazine. 645 Stewart Ave., Garden City, NY 11530, Tel. (516) 222-2500.

Electronics. 1221 Avenue of the Americas, New York, NY 10020, Tel. (212) 997-1221.

Electronics Buyers' Guide. New York· McGraw-Hill, Inc., published annually.

ISA Directory of Instrumentation. 2 vols. Pittsburgh: Instrument Society of America.

ISA Transducer Compendium. 3 vols. New York: IFI/Plenum Data Corporation, 1969.

Howard, J. L. "Error Accumulation in Thermocouple Thermometry." *Temperature,* vol. 4, part 3. Pittsburgh: ISA, 1972, p. 2017.

IEEE Transactions on Industrial Electronics and Control Instrumentation. New York: Institute of Electrical & Electronics Engineers, published quarterly.

IEEE Transactions on Instrumentation and Measurement. New York: Institute of Electrical & Electronics Engineers, published quarterly.

Instrumentation Technology (INTECH). 400 Stanwix St., Pittsburgh, PA 15222. Tel. (412) 281-3171.

Instruments & Control Systems. Chilton Way, Radnor, PA 19089, Tel. (215) 687-8200.

Manual on the Use of Thermocouples in Temperature Measurement. STP 470A. Philadelphia: American Society for Testing and Materials (ASTM), 1974.

"Measurements & Control Buyers Guide." *Measurements & Control.* Pittsburgh: Measurements & Data Corp, published in installments 6 times per year.

M & C has good tutorial material on topics in measurement and control, plus the continually updated 6-part Buyers Guide.

"Measurement Systems Short Course." Brochure. Phoenix, AZ 85018: Stein Engineering Services, Inc., 5602 East Monte Rosa, 1980.

Powell, R. L., Burns, G. W., et al. *Thermocouple Reference Tables Based on the IPTS-68.* National Bureau of Standards Monograph 125, 410 pp. Washington, DC: U.S. Department of Commerce, 1974.

Publications & Education Aids 1979 (or latest). Catalog. Pittsburgh, PA: Instrument Society of America.

Sigmann, R. T., ed. *PRÉCIS Abstracts of Electronics Publications.* Ridgewood, NJ 07450: Technical Information Distribution Service, monthly.

Standard Temperature-Electromotive Force (EMF) Tables for Thermocouples.

ASTM # E230-72. American National Standard C96.2-1973. Reprinted from the *Annual Book of ASTM Standards.* Philadelphia: American Society for Testing and Materials, 1973.

Weighing and Measurement. Rockford, IL: Scale Journal Publishing Co., published bimonthly.

MAGAZINE ARTICLES

Coping with Interference

Brokaw, A. P. "Designing Sensitive Circuits? Don't take grounds for granted." *EDN,* 5 October, 1975, p. 44ff.

Dreger, Donald R. "Plastics that Stop EMI." *Machine Design,* 26 July, 1979, pp. 114–19.

McDermott, Jim, "EMI Shielding and Protective Components." *EDN,* 5 Sept., 1979, pp. 165–76.

Morrison, Ralph. "Answers to Grounding and Shielding Problems." *Instruments & Control Systems,* June, 1979, pp. 35–38.

Patstone, W. "Designing Femtoampere Circuits Requires Special Considerations." *EDN,* July 1, 1972.

Severinsen, John. "Up Your EMI Protection." *Instruments & Control Systems,* November, 1979, pp. 43–46.

Temperature Measurement

Hall, John. "The Highs and Lows of Temperature Monitoring." *Instruments & Control Systems,* June, 1979, pp. 25–31.

Lovuola, Victor J. "Preventing Noise in Grounded Thermocouple Measurements." *Instruments & Control Systems,* January, 1980, pp. 31–34.

Masek, J. M. "Guide to Thermowells." *Instruments & Control Systems,* April, 1979, pp. 39–43.

Mendelssohn, Alex. "Temperature Transducers: What's Cooking?" *Electronic Products Magazine,* May, 1979, pp. 51–58.

Scharlack, R. "Simplify and Improve Heat Control by Combining Temperature Sensor and Heater." *Electronic Design,* 20 October, 1979, pp. 106–110.

Timko, M. P. "A Two-Terminal Temperature Transducer." *IEEE Journal of Solid-State Circuits.* vol. SC-11, no. 6, December, 1976, pp. 21ff.

Williams, J. "Designer's Guide to Temperature Control" *EDN*, 20 June, 1977.
—— "Designer's Guide to Temperature Measurement." *EDN,* 20 May, 1977.
—— "Designer's Guide to Temperature Sensors." *EDN,* 5 May, 1977.

Pressure Measurement

Bicking, Robert E. "Positive Feedback Compensates Tempco of Pressure-Transducer Sensitivity." *Electronic Design 7.* 29 March, 1979, p. 148.

Hall, J. "Monitoring Pressure with Newer Technologies." *Instruments & Control Systems*, April, 1979, pp. 23–29.

Slomiana, Maria. "Selecting Differential-Pressure Instrumentation." *Instrumentation Technology,* August, 1979, pp. 32–41.

Williams, A. Wayne. "Pressure Transducer Selection is Never Simple." *Electronic Products Magazine,* September, 1978, pp. 79–89.

Force Measurement

Elengo, J., Jr., "Applying Weight-Sensing Transducers." *Machine Design,* 23 August, 1979, pp. 82–86.

Orlacchio, A. W. "Shock and Vibration Transducers." *Electronic Design,* 11 October, 1974. (Available from Gulton Industries, Costa Mesa, CA).

Williams, J. "This 30ppm Scale Proves that Analog Designs Aren't Dead Yet." *EDN*, 5 October, 1976.

Flow Measurement

Morris, H. "What's Available in Ultrasonic Flowmeters." *Control Engineering,* August, 1979, pp. 41–45.

Scoseria, J. P. "Cupless Anemometer Has Diode Wind-Sensor." *Electronics,* 30 August, 1979, p. 156.

Level Measurement

Hall, John. "Level Monitoring—Simple or Complex?" *Instruments & Control Systems,* October, 1979, pp. 215–220.

Signal Conditioning

Brown, Cy. "Applications for Instrumentation Amplifiers." *Electronic Products,* 17 May, 1971.

DiRocco, J. V. "Signal Conditioning for Analog-toDigital Conversion in Instrumentation Systems." *Electronic Instruments Digest* (Kiver Publications), May, 1976.

Harte, J., Jr. "Analog Panel Meters—Alive and Well!" *Instruments & Control Systems,* July, 1979, pp. 19–23.

Haun, Alan. "The Truth About Isolation Amplifiers." *Electronic Products Magazine,* October, 1979, pp. 57–61.

Polakowski, J., Burham, F., Westwood, W. "Choosing the Right Resistors: Three Experts Give their Views." *Electronic Design,* 19 July, 1979, pp. 74–78.

Teschler, L. "Transducers for Digital Systems." *Machine Design,* 12 July, 1979, pp. 64–75.

Zicko, C. Peter. "New Applications Open Up for the Versatile Isolation Amplifier." *Electronics,* 27 March, 1972, pp. 22ff.

DEVICE INDEX

GENERAL INDEX